DATA HANDLING FOR SCIENCE AND TECHNOLOGY

DATA HANDLING FOR SCIENCE AND TECHNOLOGY

An Overview and Sourcebook

Edited by

STEPHEN A. ROSSMASSLER
National Bureau of Standards,
Washington, U.S.A.

and

DAVID G. WATSON
University Chemical Laboratory,
Cambridge, England

Sponsored by
CODATA and UNESCO

1980

NORTH-HOLLAND PUBLISHING COMPANY
AMSTERDAM · NEW YORK · OXFORD

ISBN: 0444 86012 6

Published by:
North-Holland Publishing Company
P.O. Box 103
1000 AC Amsterdam, The Netherlands

for:

Committee on Data for Science and Technology,
51 Boulevard de Montmorency, 75016 Paris, France

and

United Nations Educational, Scientific and Cultural Organization,
Place Fontenoy, 75700 Paris, France.

Sole distributors for the U.S.A. and Canada:
Elsevier North-Holland, Inc.
52 Vanderbilt Avenue,
New York, N.Y. 10017
U.S.A.

Library of Congress Cataloging in Publication Data

Main entry under title:

Data handling for science and technology.

Bibliography: p.
Includes index.
1. Information storage and retrieval systems--Science--Handbooks, manuals, etc. 2. Information storage and retrieval systems--Technology--Handbooks, manuals, etc. I. Rossmassler, Stephen A. II. Watson, David G. III. International Council of Scientific Unions. Committee on Data for Science and Technology. IV. United Nations Educational, Scientific and Cultural Organization.
Z699.5.S3D36 025'.065 80-16219
ISBN 0-444-86012-6

Printed in the Netherlands

FOREWORD

This book is concerned with publishing numerical scientific data, gaining access to it, and ascertaining its reliability. The intended audience includes "...people who generate, publish, abstract, collect, evaluate, repackage, disseminate and apply data...or are involved in providing training courses in the handling of data...or in administering funding for all these activities." A definitive treatment of each of these aspects would not have been feasible; instead, a guidebook -- a set of signposts -- is being offered to point the way through the extensive bibliographies appended to the individual chapters.

The book has been developed as a joint enterprise between CODATA (The Committee on Data for Science and Technology established by the International Council of Scientific Unions) and Unesco within the UNISIST framework and the Unesco General Information Programme.

CODATA's prime concern is "to improve the quality, reliability, or accessibility of data of importance to science and technology." The aim of the Unesco General Information Programme is "development and promotion of information systems and services at the national, regional and international levels." The UNISIST programme, now incorporated within the General Information Programme, aims at the ultimate establishment of a voluntary network of the world's information systems and services in order to facilitate information transfer to the international scientific communities. This joint endeavour thus represents a step toward the achievement of goals for both organizations.

The need for such a volume was perceived by Professor Masao Kotani (President of the Science University of Tokyo, then Chairman of the CODATA Task Group on Accessibility and Dissemination of Data and a Japanese delegate on the UNISIST Steering Committee, and now President of CODATA). With support granted by Unesco to CODATA the authors of the various chapters and the editorial team of Drs. S. A. Rossmassler and D. G. Watson expeditiously assembled the text.

It has been observed that "A scientific document is a package of information containing data, analysis, and interpretation in varying proportions. As the document becomes older, the value of the data will be enhanced relative to that of the analysis and interpretation -- except to the historian --," [R. N. Jones, cited in UNISIST Advisory Committee minutes, (1976)] hence the importance of developing techniques for data handling and management. In the physical sciences the data are usually sufficiently well defined to be clearly separable from interpretation, but in the biological and geological sciences the distinction between data and interpretation is, in some instances, far more difficult. For this reason special attention has been given to data handling on the bio- and geo-sciences.

The book addresses itself, therefore, to producers and users of data (and normally a scientist is both) as well as to information specialists increasingly called upon to provide numerical data to scientific users. It should prove invaluable in scientific libraries and other information centres in both developed and developing countries. It is a book which by its very nature is timely and necessary, and yet will clearly require periodic revision since it is concerned with a rapidly developing field.

Edgar F. WESTRUM, Jr.
Secretary General of CODATA

Adam WYSOCKI
Formerly Director,
Division of the General Information
Programme
UNESCO

EDITORIAL NOTE

The papers and discussions incorporated in this volume have been slightly edited to the extent considered necessary for the reader's assistance. The opinions expressed and the general style adopted remain, however, the responsibility of the named authors. The opinions expressed are not necessarily those of Unesco or of CODATA and do not commit the Organizations. Authors are themselves responsible for obtaining the necessary permission to reproduce copyright material from other sources.

For the sake of speed of publication the contents have been printed by composition typing and photo-offset lithography. Within the limitations imposed by this method, every effort has been made to maintain a high editorial standard; in particular, the units, symbols, and conventions employed are to the most practicable extent those standardized or recommended by the competent international scientific bodies.

The designations employed and the presentation of material throughout the book do not imply the expression of any opinion whatsoever on the part of Unesco or of CODATA concerning the legal status of any country, territory, city or area or of its authorities, or concerning the determination of its frontiers or boundaries. The mention of specific companies or of their products or brand-names does not imply any endorsement or recommendation on the part of CODATA or Unesco.

PREFACE

A comprehensive treatment in book form of the problems and techniques of handling numerical data was first suggested in 1975 by Professor Masao Kotani (at the time Chairman of the CODATA Task Group on Accessibility and Dissemination of Data) as a part of the long-range plan of CODATA. When the task was offered to the co-editors of the present volume, we felt it would be far too time-consuming an undertaking. We proposed instead a more modest approach to the subject, in the form of a sourcebook which would introduce the reader to the literature on the various aspects of numerical data handling, and offer some guidance through brief comments and interpretive notes.

The sourcebook concept survived, but it soon became clear that the comments and interpretation required expansion into a more systematic analysis. Thus the final product has returned at least part-way toward the original suggestion. We wish to emphasize, however, that the coverage is not comprehensive or authoritative. The book is, as the title indicates, an overview and a guide to sources in the original literature where the student can learn far more than we could compress into a single volume.

Because of the international nature of the sponsoring organizations, we took pains to enlist chapter authors from a number of countries. We are pleased that one result is the expression of a wide range of viewpoints and enough overlap among chapters to permit different contributors to emphasize different aspects of the same problem. There is a modest price to pay for this variety -- a lack of uniformity in style from one chapter to the next. We trust that no one will be offended. We have provided a considerable degree of cross-referencing between chapters, and much duplication in the references, since availability of any one source cannot be assured in all countries and regions of the world.

We feel that a book of this type cannot remain up-to-date for long, since the subject it treats is changing rapidly. We recognize that a second attempt to address this subject should probably cover the ground more thoroughly and in a more homogeneous manner. In spite of the shortcomings of the present work, we hope that those who create, manipulate, or apply numerical data will find here some useful insights into how their own efforts relate to other parts of this important whole.

In Memoriam

STEPHEN A. ROSSMASSLER

Dr. Stephen A. Rossmassler died suddenly on August 14, 1979, just as this book was going to press. His untimely death represents a great loss for the community symbolized by CODATA. He was active in the field of data evaluation for more than 15 years and played a major role in the development of the National Standard Reference Data System in the United States. As a charter member of the CODATA Task Group on Accessibility and Dissemination of Data, he helped formulate a meaningful program on data dissemination and gave generously of his energies to the various activities of the Task Group. His scholarly approach to complex problems, his conscientious service to many organizations, and his urbane but gentle wit will be remembered by all. This volume is dedicated to his memory.

LIST OF CONTRIBUTORS

HEINZ BARTELS is a physiologist presently head of the Department of Physiology, Hannover Medical School. His main research fields are medical and comparative physiology of respiration, blood gas transport, and perinatal physiology. He has contributed data for many Biological Handbooks (FASEB) since 1953. Mailing address: Department of Physiology. Medizinische Hochschule, D-3000 Hannover 61, Federal Republic of Germany.

PATRICIA W. BERGER is a librarian and presently Chief of the Library and Information Services Division, National Bureau of Standards. Her major concerns include continuing education for information professionals and paraprofessionals and automation of information systems. Mailing address: Library and Information Services Division, National Bureau of Standards, Washington, D.C. 20234.

ISAAC ELIEZER is a physical-inorganic chemist presently Manager of Energy Research and Professor of Chemistry at Montana State University. His present interests include information processing, computer modelling and simulation, energy conversion, molecular structure, thermodynamics, and high-temperature science. He is a member of the CODATA Task Group on Accessibility and Dissemination of Data, the IUPAC Subcommission on Solubility Data, the Advisory Committee of the Montana Energy Research and Development Institute, and the Editorial Board of the Solubility Series of Pergamon Press. He is a Fellow of the Royal Institute of Chemistry and a member of the Society of Sigma Xi. He has published nearly a hundred articles on chemistry, science information, and science education. Mailing address: Department of Chemistry and Office of MHD and Energy Research, Montana State University, Bozeman, Montana 59717 U.S.A.

URSULA HAUSEN is a medical librarian presently head of reference and literature research department, Library, Hannover Medical School. She is concerned with user education in medical libraries. Mailing address: Bibliothek. Auskunfts- und Informationsabt., Medizinische Hochschule, D-3000 Hannover 61, Federal Republic of Germany.

KOZO ISHIGURO is an optical physicist presently supervising practical laboratory work of undergraduate students as a member of the faculty of science and technology, Science University of Tokyo. He is secretary of the physics division of the technical vocabulary committee and cooperating member for developing courses in the study of science of the Ministry of Education, Japan. Mailing address: Science University of Tokyo, Kagurazaka 1-3, Shinjuku-ku, Tokyo, 162, Japan.

G. DORIEN JAMES is a nuclear physicist in charge of neutron time-of-flight experiments on the Harwell synchrocyclotron and is engaged in the measurement and analysis of neutron total and fission cross-sections. His work has established the presence of sub-threshold resonances--caused by nuclear shape isomers--in the fission cross-sections of several nuclei, and he has pioneered the use of non-parametric statistical methods as a means of revealing data structure attributable to nuclear shape isomerism. He coordinates the work of an International Nuclear Data Committee sub-group on neutron energy standards. Mailing address: Nuclear Physics Division, AERE Harwell, Oxfordshire OX11 ORA, England.

MAKOTO KIZAWA is a professor at the Department of Information and Computer Sciences, Faculty of Engineering Science, Osaka University, Japan. He was an electrical engineer and later an information scientist at Electrotechnical Laboratory, Tokyo, from 1948 to 1970. His recent works have been concerned mainly with information retrieval. He was a member of the CODATA Task Group on Computer Use until 1976, and has been the secretary of the CODATA Task Group on Accessibility and Dissemination of Data since 1972. Mailing address: Faculty of Engineering Science, Osaka University, Toyonaka, 560, Japan.

MASAO KOTANI is a theoretical physicist, and also interested in molecular biophysics and problems of scientific information. He is professor emeritus of the University of Tokyo, and a member of the Japan Academy. Since 1970 he has been President of the Science University of Tokyo. He has served CODATA as national delegate, Bureau member, and as chairman of the Task Group on Accessibility and Dissemination of Data, and since May 1978 has been President of CODATA. Mailing address: Science University of Tokyo, Kagurazaka 1-3, Shinjuku-ku, Tokyo, 162, Japan.

RITA G. LERNER is a chemical physicist, presently Manager, Special Projects for the American Institute of Physics. She has been involved in a variety of projects dealing with the development of new information services in physics, and was Project Director on a grant from the National Science Foundation for "Data Tagging and Flagging in the Physical Sciences." Mailing address: American Institute of Physics, 335 East 45th Street, New York, N.Y. 10017 U.S.A.

STEPHEN A. ROSSMASSLER is a physical chemist serving as a Program Manager in the Office of Standard Reference Data, National Bureau of Standards. He manages a group of data evaluation projects and data centers concerned with performance and durability of solid materials. He is a member of the CODATA Task Group on Accessibility and Dissemination of Data. Mailing address: Office of Standard Reference Data, National Bureau of Standards, U.S. Department of Commerce, Washington, D.C. 20234 U.S.A.

ALAN H. SHAPLEY is Director of the National Geophysical and Solar Terrestrial Data Center of the U.S. National Oceanic and Atmospheric Administration. He has been active in ionospheric and solar-terrestrial research and in data management. He is a coopted member of CODATA, representing the ICSU Panel on World Data Centres, one of his many involvements in ICSU activities since and before the International Geophysical Year. Mailing address: D6, NOAA, Boulder, Colorado 80303 U.S.A.

R. TOMLINSON is a geographer and is Chairman of the Commission on Geographical Data Sensing and Processing of the International Geographical Union. He is also Chairman of the CODATA Task Group on Methods for Handling Space and Time Dependent Data. He is President of Tomlinson Associates, a firm of consulting geographers that advises governments on the interactions between decision-making processes, geographical data handling techniques, and methods of spatial analysis.

JANE C. TUCKER is a librarian and computer systems analyst, presently Systems Librarian in the National Bureau of Standards Library. She analyzes Library programs in support of the scientific and technical work of the Bureau. She is the NBS alternate delegate to the American National Standards Committee Z39: Library Work, Documentation and Related Publishing Practices. Mailing address: National Bureau of Standards Library, Washington, D.C. 20234 U.S.A.

DAVID G. WATSON is a chemical crystallographer presently Assistant Director of the Crystallographic Data Centre at the University of Cambridge, England. His work is concerned with the critical evaluation, compilation and dissemination of crystallographic data on organic and organometallic compounds. He is chairman of the Data Commission of the International Union of Crystallography and chairman of the CODATA Task Group on the Accessibility and Dissemination of Data. Mailing address: University Chemical Laboratory, Lensfield Road, Cambridge CB2 1EW, England.

ACKNOWLEDGMENTS

All of the co-authors and editors wish to express their appreciation to the officers of CODATA for their encouragement to take on the responsibility of preparing this book, and to the members of the staff of the CODATA office for their support. In particular Edgar F. Westrum, Jr., the Secretary General of CODATA took an active role in the month-by-month work, and personally guided the preparation of the final text. We cannot thank him enough. Three prized co-workers were Jacquelyn Hall at the University of Michigan, who prepared the camera-ready copy, Mary E. Schlager at the National Bureau of Standards and Kerstin Watson at Cambridge University, who typed, revised, retyped, reformatted, and reretyped. Gertrude B. Sherwood and Jeanne R. Bride, also of the National Bureau of Standards, provided heroic proofreading services, Cynthia Goldman of the same institution tirelessly tracked down references, and Ruthe A. Pendergrass at the University of Michigan provided final revisions and keyboarding. We are truly grateful for their judgment, skill, and patience.

TABLE OF CONTENTS

Data Handling for Science and Technology
S.A. Rossmassler and D.G. Watson (eds.)
North-Holland Publishing Company

INTRODUCTION

S. A. Rossmassler

This book is intended to provide an introductory survey of the basic aspects of handling scientific and technical data, and to indicate to the reader selected sources from which more details can be obtained. It is in this sense that the title carries the name "Sourcebook." The text is addressed to a varied body of users, including people who generate, publish, abstract, collect, evaluate, repackage, disseminate, and apply data, as well as those who provide training courses in the handling of data, and those who administer the funding for all these activities.

The reader will find in much of the text which follows a rather strong emphasis on the physical sciences (physics and chemistry). Several factors have contributed to this orientation: the long-term historical concern for quantitative measurement in physics and chemistry, the fact that most measurement (in all branches of science) is based on physical and chemical methods, the original focus of CODATA on these sciences, and the parallel professional background of members of the CODATA Task Group on Accessibility and Dissemination of Data (CODATA/ADD) who wrote most of the chapters. In order to mitigate this bias, one chapter has been devoted to special aspects of data handling related to the biosciences and the geosciences.

Data, as usually understood in physics and chemistry, are numerical representations of the magnitudes of various quantities. If we further include basic qualitative data (i.e., specific, but nonnumerical scientific facts) such as the chemical and geometrical structures of molecules, decay schemes of unstable nuclides, sequences of genes in chromosomes, etc., it is not unrealistic to say that "data" make up a substantial and important fraction of scientific knowledge.

The various categories of data, with indications of relevance to the different fields of science and technology, have been described in a CODATA report prepared by the Task Group on Accessibility and Dissemination of Data (1).* Table 1 is taken from that reference, and the Table of Contents of the report is provided in Appendix III.

Data handling is taken to include all of the steps of intellectual and physical manipulation involved in recording the results of observation (in laboratory, observatory, or field), interpreting and refining those results, reporting them, publishing and disseminating the report, improving accessibility to that report, re-

---

*References throughout this work are identified by arabic numbers in parentheses, usually without the antecedent "Reference" or "Ref." The references are listed at the end of each chapter. Some chapters also provide further lists (bibliographies) of source material. When numbered, the latter are identified by the letter "B" prefixed to the number, as "B-12."

TABLE 1. Varieties of Categories of Data*

	Categories of Data	Chemistry/Physics	Geo-/Astrosciences	Biosciences
a_1	Data which can be measured repeatedly	Most data	Geol. structures, rocks Accel. due to gravity Fixed stars	Most data
a_2	Data which can be measured only once		Volcanic eruptions Solar flares, novae	Rare specimens Fossils
b_1	Location-independent	Most data	Minerals Global tectonics	Most data, excluding extraterrestrial
b_2	Location-dependent		Rocks, fossils Astronomical data Meteorological data	Rare specimens Fossils
c_1	Primary observational or experimental data	Optical spectra Crystallographic F-values	Seismographic records Weather charts	Physiological data (e.g., respiration rates, blood volumes, etc.) Biochemical data (e.g., composition of tissues and organs)
c_2	Combinations of primary data with the aid of a theoretical model	Fundamental constants Crystal structures	Fossil zoning Temperature distribution in Sun	Genetic code Body surface area Model of vascular bed Dimensions of tracheo-bronchial tree
c_3	Data derived by theoretical calculation	Molecular properties calculated by quantum mechanics	Solar eclipses predicted by celestial mechanics	Prediction of phenotypic expression from genotypes

d_1	Determinable data	Most macroscopic data	Elements of planetary orbits	Gene loci Chromosome numbers
d_2	Stochastic data	Polymer data Structure-sensitive properties	Soil and rock composition Solar flares Frequency of visible meteors per unit interval	Most data
e_1	Quantitative data	Most data	Seismic data Meteorological data	Physiological data Biochemical data
e_2	Semiquantitative data	Mohs hardness scale	Wind force scale	
e_3	Qualitative data	Chemical structural formulae Properties of nuclides	Rock classification Classification of stellar spectra Fossil shapes	Amino acid sequences Taxonomic classification of organisms
f_1	Data presented as numerical values		Meteorological data	Physiological data Biochemical data
f_2	Data presented as graphs or models	Phase diagrams Stereoscopic molecular diagrams Molecular models	Geological maps Weather maps Sky mapping at a particular radio frequency (e.g., 21 cm)	Metabolic pathways Electrocardiograms Electroencephalograms
f_3	Symbolic data		Lithology in bore hole data	

Note: A given group of data can be categorized simultaneously by several 'facets' a, b, c etc.; for instance, the nature of meteorological data characterized as a_2, b_2, c_2, d_2, e_1, and f_1 (or f_2).

*Adapted from ***CODATA Bulletin*** No. **16.**

analyzing and evaluating the results where necessary, compiling them, and delivering the data to a user for final application in solving some problem. This book attempts to treat data handling as relevant to the physical, geophysical, and biological sciences, plus related engineering disciplines. It does not discuss data in the socioeconomic sciences, medicine, or the humanities.

This book attempts to discuss in the chapters which follow the several data-handling phases mentioned in the previous paragraph. The treatment is not exhaustive. Instead, a brief survey is given of the particular step being covered, with frequent reference to more specialized sources in which a specific topic is addressed in more detail. It is hoped that such coverage will permit the reader to gain a general familiarity with all steps of data handling, and to pursue in depth only those steps in which he is interested. The references, therefore, are an essential portion of this **Sourcebook.**

Since the intended readers are part of the worldwide community of people involved in science, engineering, and technology, an international range of sources has been provided. This inclusion makes for much duplication of subject matter among the sources cited, but will hopefully make it easier for the user to carry on his own studies. Multiple references, especially with differing national focus, have been provided wherever available.

Immediately following this Introduction, there is an examination of how and why data are generated, with a discussion of the differing degrees to which data have an "absolute" aspect (in the sense that they can be determined independently of any environmental conditions or measurement technique), and a brief overview of systems for measuring and recording data. A two-part chapter is devoted to data-related aspects of the biosciences and the geosciences. Next, there is a survey of statistical analysis and interpretation of data. The following four chapters examine four important systematic aspects of data handling: presentation in the primary literature, means of finding data in the primary literature, the critical operations of compilation and evaluation of data, and the standards and guidelines which give structure to data handling. A chapter is devoted to a brief overview of how computers are used in laboratory science to handle data. The final chapter offers a cross-cut analysis of the formal and informal mechanisms by which data are disseminated, involving the interactions between suppliers and users of data.

It has not been possible, in the preparation of the present text, to include a comprehensive survey of data-handling activities as they are carried out in different parts of the world. Interested readers, and those with specialized knowledge about particular practices which are used to serve the needs and circumstances of individual countries and regions, are urged to communicate such knowledge to CODATA with inclusion of appropriate references to already-published descriptions. One particularly useful and comprehensive source for the indentification of further reading is the Proceedings (2) of a conference, held in 1975, on many aspects of data handling. The reference at the end of this chapter not only identifies the proceedings, but indicates the range of topics covered by re-citing several of the items in the bibliography of the Proceedings. This book has been written, on a chapter-by-chapter basis, by authors from several countries. Some of the chapters themselves are the result of international collaboration. All of the members of the CODATA Task Group on the Accessibility and Dissemination of Data have augmented the list of citations under "Recommended Reading" in each chapter.

This book has been sponsored by two organizations, Unesco and CODATA. Unesco, the United Nations Educational, Scientific and Cultural Organization, is the promulgator of UNISIST, a part of the intergovernmental General Information Programme of Unesco. The intent of the UNISIST programme is set forth in a synopsis (3) published by Unesco in 1971 and the Unesco General Information Programme is described in Appendix II. Additional details and addresses concerning international and national data programs are provided in Appendix I.

CODATA, the Committee on Data for Science and Technology, was established by the International Council of Scientific Unions (ICSU) in 1966 to promote and encourage on a world wide basis the production and distribution of collections of critically selected numerical values of properties of substances of importance to science and technology. Originally conceived to concentrate on physics and chemistry, it has broadened its scope to include non-numeric data and time- and space-dependent data in the biosciences and goesciences as well and is especially concerned with data of interdisciplinary significance.

CODATA is governed by a General Assembly with delegates from sixteen member countries and fifteen ICSU Scientific Unions. The General Assembly meets biennially. The general focus of CODATA is described in Reference 4. The membership of CODATA is given in Appendix III, along with a brief description of its biennial conferences and reproductions of the Tables of Contents of the *Proceedings* of the 1972, 1974, 1976, and 1978 Conferences.

REFERENCES

1. *Study on the problems of accessibility and dissemination of data for science and technology.* UNISIST Report SC.74/WS/16, prepared under Unesco contract by Accessibility and Dissemination of Data (ADD) Task Group of the Committee on Data for Science and Technology (CODATA) of ICSU. Available from Unesco. Available as ***CODATA Bulletin*** No. **16**, (October 1975).

2. Gaus, W. and R. Henzler, eds. *Data Documentation: Some Principles and Applications in Science and Industry.* Verlag Dokumentation, Munich (1977).

Illustration of the scope of this Conference, and of its possible utility to readers of the present text, is provided by the following scattered excerpts from the Bibliography of the above proceedings:

- General Views, Compendia, Directories

"Gesamtverzeichnis zum Thema Faktendokumentation, Datenzentren, Datenbank." *Rechentechnik, Datenverarbeitung* **5**, Beiheft 2, 57-63 (1969).

Taraboi, V., "Concerns and achievements in the area of spreading information regarding industrial proceedings and products." Report presented at the Seminar of the UN Economic Committee for Europe Bukarest. *Probl. de informare si documentare, Bukarest* ***12***, 5-13 (1972).

- Basic Software Design of Data Bases

Dreckmann, K.-H. and G. Hofmann, "Ein Programmsystem zur Erfassung von Daten aus komplex strukturierten Tabellen." *Gessellschaft f. Informatik, 5th annual meeting* (1975).

Skronn, H. J., "Methoden der Strukturierung von Datenbanken." *Angew. Informatik* ***15***, 204-210 (1973).

- Acquisition of Data Including Quality Control

Heu, R., J. Otten, and H.-D. Purps, "Ein autarkes Datenverarbeitungs-Subsystem für das klinisch-chemische Laboratorium." *Röntgenstrahlen* 25-31 (1974).

- Management and Organizational Aspects

DATUM: Siemens; Stadt Köln: Datenverarbeitung für die kommunale Planung. Zent.stelle f. Atomkernenerg.-Dokum. (ZAED), Leopoldshafen (1974). Bundesministerium für Forschung und Technologie. *Röntgenstrahlen* FB DV 74-03.

Vigler, G. J. "Beurteilungskriterien für Datenbanksysteme." *BTA* ***12***, 290-295 (1971).

- Confidentiality and Security of Data

Assakul, K. and C. H. Proctor. Testing independence in two-way contingency tables with data subject to misclassification. *Psychometrika* ***32***, 67-76 (1967).

Nargundkar, M. S. and W. Saveland "Random-rounding: a means of preventing disclosure of information about individual respondents in aggregate data." *Statistics Canada*, Ottawa, (1972).

- Information Retrieval

Egorova, N. A., A functional predicate language for factographic data (translat. from Russ.). *Autom. docum. math. linguist.* **9**, 63-67 (1975).

Krieg, A. F., J. B. Henry, and S. M. Stratakos "Analysis of clinical pathology data by means of a user-oriented on-line data system." In Enslein, K. ed.: *Data acquisition and processing in biology and medicine*. Pergamon Press, Oxford pp. 163-172 (1968).

- Information Analysis, Evaluation of Data

Klemm, W. CODATA: Internationale Koordination naturwissenschaftlicher und technischer Datenzentren. *Umschau* p. 404 (1968).

Wilhelmi, O., Jr. and L. Brown "The information analysis center: Key to better use of the information resource." *J. Chem. Doc.* **8**, 106-109 (1968).

- Presentation of Existing Data Banks

Thomas, U. "Datenbanken in der öffentlichen Verwaltung." Asgard-Verlag, Bonn (1974). OECD-Informatik Studien 1.

Dubois, J. E., D. Laurent, and H. Viellard "DARC-System-description by concentric propagation from limited environment and modular and nulinear method of representing a symmetrical graph." *C. R. Acad. Sciences*, ***A 270***, 228 (1970).

Nemoda, L. "Aspects of the organization and operation of the central data and information bank of the National Institute for Scientific and Technical Information and Documentation (INID)." *Probl. de informare si documentare, Bukarest* ***7***, 730-755 (1973).

Synopsis of the Study Report on the feasibility of a World Science Information System ("UNISIST feasibility study") by the United Nations Educational, Scientific and Cultural Organization, and the International Council of Scientific Unions.

CODATA/ICSU (A Descriptive Pamphlet), CODATA Secretariat Paris (1977).

Data Handling for Science and Technology
S.A. Rossmassler and D.G. Watson (eds.)
North-Holland Publishing Company

DATA GENERATION

M. Kotani

ABSTRACT

This chapter examines first the reasons why data are generated, the forms in which they are recorded, and systematic field collection of data. The character, significance, and methodology of acquiring data in different disciplines are then treated; data in physics and chemistry are reviewed first, proceeding from the macroscopic level to the sub-nuclear, and concluding with the fundamental constants, with comments on the increasing levels of "absolute" character of the data. In the discussion of bio-sciences and geo-sciences data, parallels are drawn as to the "absolute" character of these data as well. Generation of data is discussed as a concept and through simple examples. The nature of measurement systems, "passive," "active," "open," and "closed" is noted, as is the need for reduction of data to standard values. Finally the methods of recording data are examined, with coverage of digitization plus digital-to-analogue and analogue-to-digital conversion requirements.

GENERAL*

Data in science and technology are usually the results of experiments or observations carried out by research workers (1-6). In some instances the primary objective of the research is to obtain the data, but more frequently the data are generated for some other purpose, e.g., to confirm the identity of a synthesized compound, and are published only if the author deems it necessary to complete his presentation of the main results of the research. As a result some potentially valuable data are not published at all. It is, however, desirable that such data be submitted to and stored in appropriate data banks and data depositories to facilitate their later utilization.

Programs for the planned and systematic production of high quality data fall into two categories: one is data generation with standard instruments and methods (7,8) by a research organization which also carries out data compilation systematically; the other is by organizations and laboratories which collaborate in the sharing of responsibilities at various locations of the world. An example of the former is the generation by independent commercial laboratories of infrared absorption spectra, while examples of the latter are the work of the International Association for the Properties of Steam, cooperative observations in astronomy and meteorology, and the programs for international collaboration in the identification and classification of microorganisms and cell lines.

---

*This section was written by Makoto Kizawa of the Department of Information and Computer Sciences, Osaka University, Japan.

In the geo- and biosciences, data are obtainable not only in laboratories but also in the field, that is, by systematically making observations of natural phenomena or systems in the location where they occur. Such field data are characterized by the fact that the environment around the quantity to be measured cannot be artificially controlled in the measurement. Chapter 3 which follows presents a more detailed discussion of data-handling in these areas.

CHARACTER OF DATA IN DIFFERENT DISCIPLINES*

Data in Physics and Chemistry

According to Vitruvius, Archimedes was asked by Hiero, King of Syracuse, to examine whether the crown executed for him was made entirely of gold. Archimedes measured the mass of the crown, and then tried to determine its volume. It was not easy to measure the volume of a body of such a complicated form. After long efforts, a brilliant idea occurred to him: to measure the rise of the water level in the bath tub upon immersing the crown in the water. His final datum, the density of the crown, was obtained as the ratio of mass over volume. By comparing the volume of the crown with the volumes of two masses of the same weight as the crown, one of gold and the other of silver, he was able to determine that some of the gold in the crown had been replaced with silver. In this case the datum of mass alone or that of volume alone has no widely valid scientific meaning; at most they may be of some interest to historians. The densities of gold and silver are, however, widely valid and generally useful scientific data.

In physics and chemistry, most data obtained by scientists and published to meet the needs of users are data concerning intrinsic properties of a well-defined substance, such as the density of pure gold, or data concerning phenomena observable on a well-defined set of two or more substances, such as diffusion constants, contact potentials, and kinetic constants of a chemical reaction. Such data are valid irrespective of place and time, and can be measured in different laboratories.

Data of broad scientific significance in physics and chemistry are, in general, reproducible data given by measuring well-defined quantities on well-defined objects under well-defined conditions. An essential factor of defining "condition" of measurement is to give the values of parameter quantities of the environment, such as temperature, pressure, humidity, etc. Data which can be determined and stated independently of said parametric quantities may be considered as "absolute" data.

Since atoms, molecules, and atomic nuclei are fundamental constituents of matter, data concerning their structures and properties are of basic importance in science. Atomic weights are regularly revised and published by the Commission on Atomic Weights of the International Union of Pure and Applied Chemistry (IUPAC). Atomic spectroscopy has provided us with data on a large number of emission and absorption

---

*While the general plan of this **Sourcebook** includes citation of appropriate sources for greater technical detail on all topics discussed, such citations are not practical for all of the different experimental procedures employed by astronomers, biologists, chemists, engineers, geologists, metallurgists, physicists, etc. There are thousands of textbooks and review papers, and hundreds of handbooks which might be listed. A more limited number of references have been selected which are broadly useful and which themselves provide many further references. The work of Wilson (5) is particularly recommended.

lines of each atom in different stages of ionization. Energies of electronically excited states (energy levels) of these atoms and atomic ions have been determined and tabulated.

Molecular data are richer in variety and number, reflecting the enormous number of molecular species. Molecules can rotate, vibrate, and be electronically excited, and this situation reflects on the varieties of molecular spectra: microwave, far- and near-infrared spectra, visible, and ultraviolet spectra. The study of the atomic architecture of molecules yields a variety of data. Basically, structural formulae of molecules are very important data, although they are qualitative, not quantitative. Quantitatively, the geometrical structure of a molecule is described in terms of the positions of its constituent atoms, or more precisely speaking, equilibrium positions of the nuclei of the atoms concerned. In recent years, data on the geometrical structures of such large and complex molecules as proteins have been accurately measured, and these data are available.

Proceeding further in the microscopic direction we find "nuclear data." Each nucleus consists of certain numbers of protons and neutrons, say Z protons and N neutrons, so that (Z,N) define a "nuclide." In the case of an unstable nuclide, decay constants or half-lives for different modes of disintegration are important data. In addition, data concerning nuclear reactions such as neutron cross sections, which give the probabilities of nuclear reactions between a neutron and various nuclides, are of basic importance in nuclear engineering. Relative internal energies of different nuclides can be compared in terms of energies liberated as kinetic energies of participating particles (protons, neutrons, alpha-particles, etc.) in nuclear reactions.

We have surveyed very briefly physico-chemical data in the following sequence: data on individual macroscopic objects; data on pure substances; data on systems of pure substances and well-defined mixtures; molecular data, atomic data; nuclear data. In this sequence the latter items are less subject to outside influence than the former, at least in principle; and the descriptions of the conditions of the objects and of their environments become more and more simple as we proceed through the sequence. Finally we reach the domain of "fundamental constants," which have the broadest significance in physics and chemistry and are useful in many other fields of natural science. The most familiar of these are:

- m_e, m_p, m_n — Rest masses of electron, proton, and neutron
- e — Elementary charge (equal in magnitude to the electric charge of a proton, or, with opposite sign, an electron)
- h, or $\hbar = h/2\pi$ — Planck constant (contained in basic laws of quantum mechanics)
- c — Velocity of light in vacuum (contained in basic laws of electricity and magnetism, Maxwell's equations)
- k — Boltzmann constant (thermal energy corresponding to 1 Kelvin)
- N_A — Avogadro constant (number of molecules in 1 mole of pure substance; used also for other kinds of particles)

Most of these fundamental constants, except the velocity of light, c, cannot be precisely determined independently, but various combinations of them are measurable.

The CODATA Task Group on Fundamental Constants (FC) carried out a least-squares determination in 1973 of these and related constants, and the results have been published as a ***CODATA Bulletin*** (9). These values are recommended for general use by the major international and scientific organizations. It is often important to make clear which values of fundamental constants have been used in writing a scientific publication. Reference to the CODATA set, with the date indicated, is a convenient way of doing this.

While many of the data of physics and chemistry relate to a specific sample, object, or event, which is repeatable or reproducible, and should give the same values for the same environmental parameter values irrespective of where and when the measurements are made, the same is not true for the geo- and biosciences, or for many aspects of engineering. Even in physics and chemistry, data on particular individual objects or phenomena are of scientific value when a datum refers to a very rare object or phenomenon, or when the object or phenomenon is of interest to scientists although it is too complicated or too delicate to be well defined in the orthodox sense. Data on the deflection of the light path in a strong gravitational field observed on the occasion of a solar eclipse may be an example of the former; observation of a very fine, complex shape of a particular snowflake may be an example of the latter. In this case, it may be of interest to make similar observations on many samples and to try to find conditions in which snowflakes of that special shape are produced. Statistical analysis will be necessary in the course of this study. A certain property or phenomenon observed on some materials may be strongly dependent on the presence of a spurious impurity or minute defects of crystal structure (structure-sensistive data), and such "structure-sensitive data" may be regarded as examples. Data concerning fracture and crack formation in solids upon application of external force are stochastic and not reproducible, so that they must be studied on a statistical basis.

Data in Biosciences*

The recording of biological data presents problems unique to science because the objects of the observations and measurements are living organisms. In the laboratory, the variability between subjects and the variability within subjects precludes the determination of absolute values, such as are available in chemistry and physics, but nonetheless provides a statistically meaningful measure. As a further complication, in the field, some observations cannot be repeated under similar conditions because they are time-dependent or space dependent, or both. To assist life scientists in reporting measurements, the CODATA Task Group on the Presentation of Biological Data in the Primary Literature has prepared a guide covering the description of experiments and observational procedures, treatment of data derived from them, and presentation of the final numerical results (10).

A crucial part of any biological study is the identification of the organisms involved. Biologists have developed a systematic classification based on morphologic, ecologic, biochemical, or other characteristics by which similar individuals can be grouped together. This system of classifying organisms involves a hierarchy by which minor units are progressively gathered into defined groups of increasingly greater scope. The basic unit in the system is the "species," a category which not only has common attributes but whose members also have a potential capability of interbreeding. For example, all human beings, genus Homo, living in this era, are classified in the species sapiens, and are referred to scientifically as Homo sapiens.

---

*See also Chapter 3A.

For purposes of ordering data, a species may be considered to correspond roughly to a pure substance in chemistry, but the comparison is far from exact, of course. Just as a substance is characterized by its molecular weight, density, solubility, melting and boiling points, so must individuals belonging to a single species be characterized as objects of observation or measurement, by indication of sex (male or female), developmental stage (fetus, child, adult; larva, pupa, adult; etc.), size (weight, length, surface area), subspecific category (race of people, breed of animal, variety of plant), and other associated variables.

A vast fund of data has been accumulated on the properties, constituents, and functions of animals and plants. They vary from the physical properties of proteins to the growth patterns of various animals, and from the mineral constituents of plant tissues to the effects of environment and chemicals on all species. Data on these and many other subjects are available in the literature, and cover organisms from aardvark to zebra alphabetically and virus to whale dimensionally (11).

In the biosciences, the data generated are usually of a physico-chemical nature, and practically all of the measuring techniques of chemistry and physics are used. Most of the quantitative data in the life sciences are related to the fields of biochemistry and biophysics. This is particularly true, for example, at the molecular level where biological purity and chemical purity meet. In the chemical sense, a particular protein may consist of identical molecules, such as the hemoglobin in blood or the actin in muscle. Though the hemoglobins from different species may perform similar biological functions, they are more or less different in chemical structure. Even among individuals of the same species, the hemoglobin of certain small groups will differ slightly in molecular structure from that of normal subjects. This lack of strict purity of genetic information among individuals of the same species is further evidence of the inherent variability between living organisms.

The precision of chemistry and physics is based on the capability of repeating experiments in different laboratories under defined conditions. This capability became manifest when agreed-upon standards of measurement were developed, and reagent- and analytical-grade chemicals became available. Thus it became possible to determine the variables involved in a particular phenomenon, and restrict to one the number of variables being tested. Such precision in the biosciences is unattainable, but the development of inbred strains of animals made possible the use of organisms whose members have essentially the same genetic constitution. As a result of this carefully controlled inbreeding, experiments in the life sciences can be repeated using an animal strain in which the variables are known or can be regulated. Inbred animals lines can be considered the reagent-grade chemicals of the life sciences.

The high degree of purity achieved through inbreeding does not pertain to most organisms for which data are obtained. Another difference in biological and chemical purities is the effect of varying environmental factors on living things. This is true even between monovular twins who share the same genetic inheritance. The post-natal environmental influences are particularly evident in such phenotypic expressions as body height and weight. To be scientifically significant, data for such properties are usually derived by measuring a large number of individuals, and applying statistical methods appropriate to the design of the experiment.

Much of the data basic to biology is genetic in character. An example is the identification of particular genes which correspond to given phenotypes, such as eye color or hair texture; another is the number of chromosomes in cells for each species. The arrangement of various genes shown as gene maps is a basic genetic datum for each species, although known for a limited number of species at present. Each gene contains a self-reproducible record of the chemical structure of specific proteins. This record is "written" on DNA in terms of four kinds of "characters," called nucleotides, and each sequence of three nucleotides defines an amino acid constituent of the protein. The correspondence between 64 nucleotide triplets and 20 amino acids is called the genetic code, which constitutes the biochemical data fundamental to all species of animals and plants. In its broad validity, the data of the genetic code may be compared with data on fundamental constants in physics. Though genetic-code data and many other biological data are qualitative, rather than quantitative, they are used in research and application in some of the same ways as numerical data are utilized in physics. For instance, data contained in the *Biology Data Book* (11) are primarily qualitative, and even when given numerically, they usually refer to "typical cases" rather than to "standard values." Thus in the biosciences, values are generally presented as either the mean plus or minus the standard deviation, or the mean and the lower and upper limit of the range of individual values about the mean (either observed or statistical). In order to reflect the variability of living things, it is of great importance that the range, rather than the mean, be given.

Just as the "species" constitutes the basic unit in the classification of organisms, the "cell" constitutes the basic unit in the structure of the individual. Although the cell is the smallest unit of structural living matter capable of functioning independently, it is composed of even smaller elements, such as the nucleus, the mitrochondrion, and other organelles within the cell cytoplasm, all encased in a semipermeable membrane. At this microscopic level, the interrelationship between biology and chemistry becomes quite evident when the important roles of biological molecules--such as nucleic acids, proteins, lipids, hormones, vitamins, etc.--are considered.

Data in the biological sciences should be presented according to accepted criteria which permit comparison with results of observations or experiments made at different times or places. It is, therefore, essential that an adequate description of the procedures for obtaining the data include a complete identification of the organisms involved, a definition of the system studied, identification of the methods and procedures used, a description of apparatus and chemicals used, and a statement of the sensitivity and resolution achieved in the measurements.

Data in Astro- and Geosciences*

In pure physics and chemistry those data which are particular to a given object for measurement and are not applicable to other similar objects are not usually considered scientifically valuable, as mentioned earlier. In astro- and geoscience this situation becomes quite different. In astronomy, stars are classified into different groups, and they are born, grow, and decay just as biological individuals, but there seem to be no "stellar codes" which correspond to the genetic code, and an astronomical concept corresponding to biological species is difficult to obtain. Furthermore, our human time span is too short, and we can observe stars only in terms of the light which reaches us from them, some rather near to us and some very remote. Thus each star or celestial body has its own scientific significance, and data concerning it are important. Sirius, for example, gives its own data of interest for many astronomers. The extreme case is our Sun. The Sun is relatively near to us and is essential to our lives on earth. Its size and shape, its mean distance from the Earth, its mass, its rotation, are represented by important data in addition to data concerning the radiation and particles it emits, the physical and chemical properties of its surface, etc.

---

*See also Chapter 3B.

While data on the Sun are important, our Earth is a body even more essential to us. Geoscience aims at studying the Earth. Data in the geosciences must normally be accompanied by records of time and position of objects on which observations were made. This is evident if one considers meteorological data or seismological data. In some cases time is not very important. For instance, various properties of a particular specimen of rock (in a stable state) can be measured at any time in a laboratory. The specimen can be transported from one laboratory to another. However, the kind of rock as specified by its geological name, such as granite, does not fully describe the specimen and does not provide enough characterization to make it a well-defined object. It is necessary to specify the place and geological condition where the rock has been obtained. Thus the data on rocks are location-dependent. The acceleration of gravity, which has important uses in physics, is also essentially geoscience data, and is location-dependent.

Since most geo-scientific data thus fluctuate with time and/or location, it is sometimes convenient to set up "standard values," based on a certain average, which can be used as a reference point. For instance, the standard atmospheric pressure (at sea level) is defined as 101.325 kPa, and standard gravitational acceleration is 9.80665 ms^{-2}.

Ages of rocks and fossils are another kind of important geoscience data. Fossils found in a geological formation are sometimes used to determine the age of the formation, but in recent years the ratio of amounts of transformed and untransformed atoms of a radioactive element is used to determine the age absolutely. Because of this application, decay constants of various radioactive nuclides are being reexamined and more accurately determined.

MEASUREMENT AND DATA REDUCTION*

Measurement

In the fields of physics and chemistry and in their areas of application, data are usually understood to be numerical representations of magnitudes of various quantities, as stated in Chapter 1. More exactly, this is expressed (12) by:

physical quantity = numerical value × unit

Data originate usually from measurement carried out in laboratories and in the field using various measuring apparatus (5,13). Measurement in this sense is a process of extracting desired information from specimens under inspection. The information is obtained as signals, usually transmitted by some carrier and is reduced to numerical figures or analogue representation and then stored in an adequate form such as tables or graphs.

Some kinds of data are obtained directly from a single type of measurement, while other kinds are obtained indirectly by combining results of several different types of measurements. A very simple example of the former is the measurement of the distance along the road around a park. This datum is directly obtained from the reading of a tape measure. An example of the latter case is the determination of how many litres of petrol are needed for a given car to run 1 km. In this case of fuel efficiency, both the distance, L, (kilometres) which the car has run and the volume of petrol, V,(litres), which the car has spent by that time must be measured to obtain the datum of fuel efficiency,

*This section was written by Kozo Ishiguro, of the Science University of Tokyo.

L/V (km/ℓ).

As a more academic example of the latter case, let us consider measurement of Young's modulus of iron. This property is determined by taking a straight iron bar of natural length, ℓ, and cross section, S, fixing it at one end surface, and measuring its elongation, $\Delta\ell$, on applying a known force F to the other end surface. Young's modulus E, is calculated by the well-known formula:

$$E = (\ell \times F)/(\Delta\ell \times S)$$

In this case, temperature and purity of the specimen of iron are important parameters to be given with the datum to make it useful and reproducible. Atmospheric pressure is not so important, since E is not sensitive to its change, but the nature of any crystal imperfection may be important and should be described.

In such a case as Young's modulus data, only the value of E is usually reported in a formal publication, and intermediate data such as $\Delta\ell$, ℓ, S, F for each measurement are not reported but may be preserved in the private records of the researcher.

To evaluate data as to accuracy or reliability it is necessary to know the internal structure of the apparatus used and the details of the principles of the measuring methods. In most ordinary handling of data, however, one need not know all such details; the instruments may be looked upon as black boxes with input and output terminals attached. For instance, the measuring apparatus for Young's modulus mentioned above may be regarded as a black box, of which the output terminal indicates a deflection of an indicator proportional to the elongation of the bar when a known tension is applied to the bar through the input terminal.

In general, a measuring apparatus is composed of a main body which is peculiar to the apparatus plus an auxiliary instrument which displays the obtained data as mechanical deflection of an indicator or as digital figures generated electronically.

The theory of measurement is frequently discussed using technical terms of electrical communication theory (14). From this standpoint, the main body of the measuring apparatus is called a "signal transducer," and the instrument which displays the data as some type of signal is called an "output recorder" or "output indicator." Such an output recorder or indicator may be usable for universal applications, driven by electrical or mechanical signals applied to the instrument. Therefore, the signal transducer must be able to change the input signals related to some natural phenomenon to electrical or mechanical signals which can drive the output recorder or indicator.

It is generally known that transmission of signals must always be accompanied by energy flow. When the energy of the output signals of the signal transducer is supplied by the input signals only, the system is called a "passive system," and when the energy is supplied by a power source contained in the signal transducer, the system is called an "active system." In the latter case, the role of the input signals is limited to actuating or modulating the output signals, and only a small output of energy from the specimen is needed. Accordingly, reaction of the specimen to the measurement conditions can be made very small.

The passive system is usually very simple and reliable, but the output energy from the system is very small when the reaction of the specimen is to be kept small. It will need a very sensitive output indicator such as a galvanometer.

Whether passive or active, the above-mentioned measuring systems are called "open systems," responding to the input signals but also responding to irregular changes in the sensitivity of the apparatus. To eliminate this defect, a measuring method called "closed system" or "null method" has been developed. The well-known potentiometric method for measuring electric voltage and the Wheatstone bridge for measuring electrical resistance are typical examples. In these instruments, the input signal is transformed to the reading of an indicator, but at the same time a signal which deflects the indicator in the reverse direction is produced by adjustment of a "compensator" in the instrument until the reading of the indicator becomes zero. Then the magnitude of the input signal is obtained from the value needed to adjust the compensator. The mechanism of this method, when automated, is said to feed back the output signal of the indicator into the input side until the output signal of the indicator becomes zero. Because this method of "closed system using feed back" is a very convenient and reliable measuring method, it is widely used in many types of measurement.

Post-Measurement Reduction

Mathematical analysis and interpretation of data according to the theory of probability are discussed in Chapter 4 (See also references 3 and 15). Therefore, it may be sufficient here to remark that the raw data obtained by measurements are "random variables" in mathematical terms and can be treated according to the theory of probability to obtain approximations to "the most probable value," "standard deviation," and "the allowed interval of the desired confidence probability," and other well-defined statistical measures (14).

The magnitudes of raw data obtained by measurement and the most probable value or the mean value obtained from the raw data by mathematical treatment must be reduced to magnitudes corresponding to an appropriate standard unit or scale, most preferably to units (12,16) in the SI system. When the parameters governing the operation of the instrument are precisely known, this conversion is done quite easily. However, in most cases a "calibration" of the instrument is necessary. The usual method of calibration is to measure the magnitude of the output signal using a well-established specimen or "standard" specimen. (for further discussion of standard specimens, such as Standard Reference Materials, see Chapter 8). When the magnitude of the reference specimen is known to be Y by the standard unit and is X by the intrinsic scale of the instrument, measured values given in the intrinsic scale should be multiplied by the conversion factor Y/X to obtain values in standard units.

Due caution must be paid to the fact that, before the SI units were generally adopted in science, various units more or less different from SI were employed, such as the International Volt, the International Ampere, the X-ray unit (for length in crystallography), etc. The X-ray unit was based on the lattice constant of calcite (and later of sodium chloride), but the conversion factor of X-ray units to the metric system is difficult to determine very precisely because lattice constants of natural crystals are not precisely constant from sample to sample. In general, conversion factors change as measuring techniques progress. Up-to-date information on conversion factors for important quantities should always be readily available to data handlers.

In some cases, several different units may be used to represent the same quantity for practical convenience. For instance, the "energy" of a photon (as well as excitation, ionization, and binding energies of atoms and molecules) may be represented in one table by electron volts; in another table by wave number units (1/cm); in still another table by kcal/mol. The Rydberg Constant is also used as an energy unit. Confusion sometimes results because these are not energy units; rather, they are units of some quantity that is proportional to energy, and the conversion factor to a true energy given in terms of the fundamental constants. The most recent authorized version of fundamental constants should always be consulted for making conversions (9, 17).

RECORDING, DIGITIZATION AND PROCESSING OF DATA

In the process of obtaining usable or refined data, raw or intermediate data have often to be handled by the experimenters. Automatic techniques may be employed, by which the data values are recorded continuously or at intervals, often as function of time; this is particularly useful to meet the need for making very large numbers of measurements at very fast rates or for a very long period, and also for reducing the data rapidly to a form which can be used efficiently and effectively. Thus, in addition to reducing human effort, elimination of incidental errors which may be caused by humans may be expected. Acquisition of data in an earthquake or volcano eruption is a good example of the beneficial use of automated recording. The phenomenon must be measured at as many locations as possible, and can be measured only while the phenomenon is going on. Furthermore, such natural phenomena do not appear at the observer's convenience -- the apparatus must be ready to make measurements at any time.

There are several techniques of recording quantitative data now widely employed, falling into three major categories according to the method of presenting data curves: pen-and-ink (thermal-, electric discharge- and scratch-pens included), photographic, and magnetic methods. Pen-and-ink recording has the merit that the recorded chart can be studied by the human eye, but the changing rate or frequency component of the signal is limited to a low value, or the order of 100 Hz. Recording of temperature, speed of massive mechanical motion, rates of most biological phenomena, etc., are typical applications of this type. A cathode ray tube interfaced with a photographic recording system has a much higher frequency limitation, but requires cumbersome processing. This can be applied to the recording of electrical transients, high frequency vibrations, etc.

Recording on magnetic media (most frequently, but not exclusively, via computer) has many distinct advantages (18). Among other factors that make magnetic recording so useful is the reproducibility of the phenomenon-derived information as an electrical signal even with changed time-bases, so that the information can be put into automatic processing after the phenomenon is over as often as the observer requests, at the desired rate, and in the proper environment for the measurement. Three analogue recording methods--direct recording (DR), frequency modulation (FM), and pulse-duration modulation (PDM)--are available, and each of them has its own signal frequency range which is dependent on the relative speed between the magnetic head and the magnetic medium (19,20). The digital recording method is particularly useful when combined with modern digital computers, as described below (also see ref. 21). The chapter of this ***Sourcebook,*** on "Computer Handling of Data" covers many aspects of this topic in more detail.

It goes without saying that most data in science and technology are intrinsically in analogue form and are measured with analogue instruments. There is, however, a growing trend toward obtaining data in digital form, thereby facilitating the use of the extensive capabilities of modern electronic computers and telecommunication systems. Digitization of data from the analogue form provides a solution to this need.

Digitization can be performed automatically by the use of an analogue-to-digital converter (AD converter), which is sometimes incorporated in the measuring instrument. Its important parameters to be considered by users are the accuracy, the precision, and the sampling rate. The precision can be estimated by the number of bits for a piece of data. The digitizing error, which is closely related to the precision, is estimated to be within a half of the step quantity for a bit, and consequently, for example, a set of nine bits at least is necessary to ensure a digitizing error of within 0.1 per cent of the full scale. This error should be well coordinated with the inherent precision of the measuring instrument. The sampling rate, which is directly related to the time interval of the measuring points, should be selected so as not to lose any part of the data which changes significantly with time.

Digital-to-analogue conversion (DA conversion) is, on the other hand, employed when the digital data, mostly processed by computers, require presentation in a form which is readily assimilated by humans. For this purpose, a plotter, electromechanical or electronic, is a major device, the incremental plotter being one of the most popular and useful types (22).

Environmental conditions, such as date, location, temperature, humidity, and atmospheric pressure should accompany the recorded data as necessary and can be automatically recorded. Parameters of the measuring instrument, accuracy, precision, and calibration method should also be noted because they are essential factors in data evaluation.

It is recognized that, as computer technology advances, there is a growing tendency toward the automation and computerization of data handling, including the collection, reduction, storage, editing, retrieval, and dissemination of data. The ways in which computers are used, however, must be carefully examined. Although computerization can be very efficient, it does not guarantee optimum planning of an experiment, nor most effective service to the final user of the data.

REFERENCES

1. CODATA/ADD: "Study on the Problems of Accessibility and Dissemination of Data for Science and Technology." ***CODATA Bulletin*** No. **16**, October 1975.

2. Popper, K. R. *The Logic of Scientific Discovery.* Hutchinson, London (1959).

3. Brillouin, L. *Scientific Uncertainty, and Information.* Academic Press, New York and London (1964).

4. Bridgman, P. W. *The Logic of Modern Physics.* The MacMillan Company, New York (1960).

5. Wilson, E. B. *Introduction to Scientific Research.* McGraw-Hill, New York (1952).

6. Freedman, P. *The Principles of Scientific Research.* Pergamon Press, New York (1960).

7. Schaefer, C., ed. CODATA *International Compendium of Numerical Data Projects.* Springer-Verlag, Berlin (1969).

8. "CODATA Directory of Numerical Data Projects." Appearing by chapters in ***CODATA Bulletin*** Chapter 1, "Crystallography," by D. G. Watson, in ***CODATA Bulletin*** No. **24**, (June 1977).

9. CODATA Task Group on Fundamental Constants. "Recommended Consistent Values of the Fundamental Physical Constants, 1973." ***CODATA Bulletin*** No. **11**, (December 1973).

10. CODATA Task Group on the Presentation of Biological Data in the Primary Literature. "Biologists' Guide for the Presentation of Numerical Data in the Primary Literature," ***CODATA Bulletin*** No. **25**, (November 1977).

11. Altman, P. L. and D. S. Dittmer, ed. *Biology Data Book,* 2nd edition., vols I-III. Federation of American Societies for Experimental Biology, Bethesda, Maryland (1972-1974).

12. S. U. N. Commission. *Symbols, Units and Nomenclature in Physics.* Document U.I.P. **11** (S.U.N. 65-3), (1965).

13. Partington, J. R. *An Advanced Treatise on Physical Chemistry.* Longmans Green, London.

    Vol. 1. *The properties of gases.* (1949)
    Vol. 2. *The properties of liquids.* (1951)
    Vol. 3. *The properties of solids.* (1952)
    Vol. 4. *Physico-chemical optics.* (1953)
    Vol. 5. *Molecular spectra and structure, dielectrics and dipole moments.* (1954)

14. Terao, M. *Theory of Measurement* (In Japanese). Iwanami Press, (1975).

15. Cochran, W. G. and G. M. Cox. *Experimental Designs,* 2nd edition. John Wiley, New York (1957).

16. *The International System of Units (SI)*. NBS Special Publication 330 (August 1977). Washington, D. C. U.S. Government Printing Office C13.10:330/4, Stock No. 003-003-01784-1.

17. DuMond, J. W. M. "Pilgrim's Progress in Search of the Fundamental Constants." *Physics Today* 26-43 (October 1965).

18. Weber, Paul J. "Magnetic Tape Recorders." *Instruments and Control Systems* 380-383 (1959).

19. Halfhill, D. W. "Recording Techniques." Ibid., **32**, 384-388 (1959).

20. Fetty, K. "Magnetic Tape System for Analog, PDM, or FM." Ibid., **32**, 392-393 (1959).

21. Kezer, C. "The Digital Magnetic Tape Recorder." Ibid., **32**, 389-390. (1959).

22. Talley, D. "Automatic Plotting in the Third Generation." *Datamation* **13**, 22-26 (July 1967).

BIBLIOGRAPHY

B-1. Annual Reviews, Inc., 4139 El Camino Way, Palo Alto, California 94306, USA.

Each year this organization publishes *Annual Review* volumes covering selected topics in each of the following fields of science:

Anthropology
Astronomy & Astrophysics
Biochemistry
Biophysics & Bioengineering
Earth & Planetary Sciences
Ecology & Systematics
Energy (new series--1976)
Entomology
Fluid Mechanics
Genetics
Materials Science
Medicine
Microbiology
Nuclear Science
Pharmacology
Physical Chemistry
Physiology
Phytopathology
Plant Physiology
Psychology
Sociology (new series--1975)

Data Handling for Science and Technology
S.A. Rossmassler and D.G. Watson (eds.)
North-Holland Publishing Company

TREATMENT OF DATA IN THE BIOSCIENCES

H. Bartels and U. Hausen

ABSTRACT

Major challenges to the measurement, reporting, evaluation, and use of biological data arise as a result of the variability which accompanies life processes. These difficulties tend to set biological experiments and the resulting data apart from other scientific data. This chapter examines some of the methods, experimental procedures, choice of subjects, and internal phenomena which influence biological data. Special categories of data, use of computers, classification requirements, and data presentation needs are discussed. Sources of untabulated as well as compiled and tabulated data are provided.

INTRODUCTION

Life processes give rise to special requirements in the handling of biological data. Three factors commonly used to distinguish living from nonliving subject matter are reproduction, metabolism, and growth. Inherent in each of these functional activities is the variability that differentiates biological from other scientific data. Another important characteristic of biological data is the profound influence on morphology and function of organs, cells, parts of cells, and molecules when isolated from the whole living organism.

One of the most dangerous "imperatives" promulgated among life scientists is to arrange experiments on a solid scientific (i.e., physicochemical) basis without regard for other considerations. Many of the data collected under such conditions may be absolutely worthless or—at least—misleading. Two examples may serve as illustrations: i) Cardiac physiology and pathology was—and still is—influenced by the Frank-Starling law, which is deduced from dog heart-lung preparations and applied to man and other animals although it plays almost no role in normal dog or man. The extra-cardial nerves are physiologically much more important than the mechanism found in the preparation without these nerves. ii) "Purification" of biological substances can lead to molecular structures never present in a living plant or animal. Data from "stripped" chlorophyll or hemoglobin can be totally misleading if applied to a living organism.

Another consideration is that almost all data from animals and many of those from plants cannot be acquired without consciously or unconsciously disturbing normal life processes by our methods. Change of environment, the act of taking measurements, and tranquilization and anesthesia in animals usually interfere with some normal biological function. Even blood sampling or blood pressure measurements in unanesthetized man influence the data measured.

In the biosciences, it is essential to know the species from which data were obtained. The application of classification is therefore absolutely necessary. Cultivation of plants and breeding of animal strains, domestication, and readaptation of domesticated animals to their natural living conditions may be other reasons for data variation. Environmental factors play an enormous role in biology, but in many experiments the investigator is unaware of all the environmental factors influencing the data.

Biological variability is one of the major prerequisites for life. Individually the processes commonly denoted as adaptation and acclimatization produce such variations. From generation to generation, mutations and selection -- and in higher animals, learning ("social heredity"), too --are other causes of data variation.

Variations of biological data in time (biorhythms) are well known for some biological parameters, but there may be many other variations due to biorhythms that have yet to be proved. The most investigated rhythms are the circadian (diurnal) and seasonal ones.

Besides numerical data in the biosciences, nonnumerical data play an important role. Morphological data describing different sequoia species, African and Indian elephants, nuclei of white blood cells, three-dimensional molecules, the embryological development of vertebrates, and so on cannot be documented by figures or, at least, doing so would be extremely impractical.

Considering all of these aspects peculiar to biology, a young scientist may come to the conclusion that only experiments with monovular twins kept in the same environment could be performed and not much more than body weight and length could be measured. There is no valid reason for such experimental nihilism. The reverse is true; in addition to investigating the still unknown phenomena, a great variety of "known facts" should be examined and all factors possibly influencing the data should be considered.

"Massaging of data" should be given a far lower priority than the tabulation and subsequent publication of data together with careful documentation of the experimental conditions.

The aim of this contribution is to present the most common factors and conditions influencing numerical and nonnumerical data in the biosciences, and to provide available references.

METHODS

Although many of the numerical data in the field of biology were and are measured with so-called "exact" physical, chemical, or physico-chemical methods, these values tend to be far less reliable than in physics and chemistry. The reasons are manifold and will be discussed below.

Standardized methods

These yield the most reliable, comparable, and applicable data. They are generally described in great detail. Commercial instruments are usually preferred for making measurements, but the model number and manufacturer must be cited. The authors of original papers, the reviewers of articles, and the contributors of tables in handbooks should mention the methods used as exactly as possible. Nowadays most journals and data books follow this rule satisfactorily.

A great number and variety of methods are applied in the biosciences. Examples of reliable methods are measurements of length, weight, volume, pH, osmotic pressure, electrocardiography, cell count in body fluids, and concentration of slowly metabolized constituents of cells and fluids in plants and animals. However, even the multi-volume handbooks in a special field are often incomplete and partially out of date. Information taken from them should be augmented by the latest literature, if necessary. Bibliographies on methodology in the biosciences are available (1-3).

Non-standardized or only apparently standardized methods

These yield results that are not always comparable. Therefore, it is even more important to give exact references for the method used and to mention one's own modifications. Contributors to biological tables often are tortured by twinges of conscience as they attempt to take into account such methods and modifications. Frequently this struggle is documented in a multitude of footnotes indicated by every conceivable symbol. Nevertheless, many of the data generated are based on such methods and are subsequently tabulated and used by scientists. It must be emphasized, however, that these data are to be treated critically; in case of doubt, the original paper, as usual, must be consulted.

From the thousands of methods in use, some have been selected as examples of those considered to be fairly standardized and exact: measurements of biological volumes only indirectly accessible (intracellular water content, glomerular filtrate volume, residual lung volume) require corrections in order to make allowance for excretion, evaporation, absorption of tracer substance, etc. Brachial blood pressure measurements in human subjects are influenced by cuff dimensions, but the standardization of cuffs is just starting. Dimensions and weights of macromolecules can be influenced by the "purification" or fixation process. Morphological measurements on organs, tissue slices, or cells are influenced by shrinking factors during fixation. Flow measurements of liquids in plants and animals can be influenced by the cannula necessarily inserted into the fluid canal. These examples show the influence that methods may exert on the data. Required corrections cannot always be estimated with sufficient precision.

INFLUENCE OF EXPERIMENTAL PROCEDURES ON RESULTS

In a broader sense than mentioned in the preceding paragraph, experimental conditions are often associated with undesirable effects which influence the data. In animal studies, anesthesia is one of the most important factors interfering with almost all functions, but predominantly with the central nervous system, with respiration, and with circulation. Qualitatively, much is known about these influences, but a quantitative application is hazardous. Anesthetics differ in their effect on species, and the same is true for tranquilizers (4-7). Furthermore, anesthetics or tranquilizers may interfere with the action of biological substances or drugs under investigation and yield misleading results (8,9). Dosage and route of administration of all substances must be indicated. Transport of plants and animals, fixation of animals, pain, and excitation influence many functions, mainly via the autonomic nervous system. Influence of anesthesia and other restraining measures can be avoided by wireless transmission of data, telemetry, via implanted measuring devices.

The steady state concept is well defined for investigations of overall metabolism (e.g., oxygen consumption) in man and higher animals. Measurements should be taken only when input and output of energy are balanced. Unsteady states can give rise to misleading data in many investigations, from the macroscopic level of the whole plant or animal down to the microscopic level of subcellular elements.

Body fluids can change in composition and function if in contact with air during withdrawal, if anticoagulants are added, and if other procedures are used. The sampling site is another important factor in this respect. Blood composition differs, not only between arteries and veins, but also in different veins, capillaries, etc. (10,11).

Isolated organs, parts of organs, cells, and parts of cells are functionally affected by isolation from the whole organism. Only limited physiological conditions frequently exist for perfused isolated organs, tissue slices kept in "physiological" solutions, or microorganisms grown in culture media. Organs and tissue slices of warm-blooded animals are mostly investigated in solutions saturated with 95% oxygen (oxygen poisoning!) and 5% carbon dioxide. Such investigations always make compromises necessary, and the solution or medium must be selected presumably to fulfill "physiological" conditions as closely as possible (12-14).

INFLUENCE OF CHOICE OF EXPERIMENTAL PLANT OR ANIMAL

Investigators of comparative aspects of biology take great care in accurately and precisely identifying the species which they compare. Generally, in research there is a growing understanding for the necessity of mentioning genus and species or even subspecies (races in man) as exactly as possible. The generally accepted classification of both the plant and animal kingdoms should be consulted and applied (15-20). As already mentioned in the introduction, cultivating of plants and breeding of special strains of animals (cows, sheep, chickens, etc., in agriculture; dogs, and especially small rapidly-reproducing mammals for experimental purposes) make it necessary to give strain or breed and source as precisely as possible (21-22). For the selection of animals, directories of sources of laboratory animals should be consulted (23,24). Classifications of bacteria and viruses are available (25-27). The International Association of Microbiological Societies has issued a directory of cultures of microorganisms (28). Further information is given in reference 29.

In this connection, it should be emphasized that special fields in the biosciences have their preferred experimental models. In circulatory physiology the majority of data are from dog experiments. These data are applied to medicine despite the fact that dogs show some fundamental differences in their circulatory systems as compared to man. Data on liver, kidney, and other functions are mainly derived from rat studies. In pharmacology and electrophysiology, cat experiments prevail. Nerve conduction studies are usually performed on the giant axon of *Loligo* (squid). The generalizations drawn from such data are not always viewed with necessary caution, and their validity for other species may sometimes, or even frequently, be suspect (30).

SEX DIFFERENCES, REPRODUCTION, PUBERTY, GROWTH, AGE

The sex of the plant or animal investigated has to be mentioned, if possible, even though the investigator at present believes there may be no sex differences in the data in question. In all bisexual plants and animals, events during reproduction influence many functions and data measured under such conditions. Qualitative and quantitative variations of hormones are well-known occurrences. Puberty and growth are accompanied by many obvious changes. If possible, the investigator should mention these conditions, even though he may not be interested in the developmental aspect itself. In general, the same is true for age effects in all organisms even when only parts of them, down to subcellular levels, are under investigation (31,32). Morphological and functional data as well as growth data in plants and animals are well documented (3,33,34).

ENVIRONMENTAL FACTORS

Whether we define environment as the natural surroundings of investigated living organisms or parts thereof, or include artificial surroundings such as culture media, environment always has a very important influence on morphology and function of plants and animals. Temperature, radiant energy, atmospheric pressure (hypoxic effects at altitude), water pressure (e.g., deep sea animals and plants, diving fishes and mammals), water and solutes (fresh and sea water), and "physiological" solutions (Ringer, etc.) are important environmental factors. Bioclimatology comprises many of these parameters; this is also true for items mentioned in the next two sections. Many data on the effects of varying environmental factors are documented (35-38).

Radio-frequency radiation, polarized light and ionizing radiation, sound, vibration, impact, acceleration, and influence of pollutants are becoming increasingly important because of their influence on morphology and function of biological material. Data on toxicology (39) of herbicides (40), pesticides (41), and other harmful agents (42,43) are of importance in medicine, veterinary medicine, and agriculture.

BIOLOGICAL RHYTHMS

Knowledge of biological rhythms has been increasing exponentially in recent years. Circadian (diurnal) and seasonal rhythms, dependent on light and temperature changes, influence a great number of morphological and functional parameters. Many of the results have been tabulated, and new results are currently being published. Data on tidal and lunar rhythms and electrostatic, magnetic, and gamma-radiation effects on biological rhythms are available (44,45). Even the investigator to whom biological rhythms are not of primary interest must take them into account if only for their effects on his own research efforts.

In a broader sense, frequency of the heart beat and breathing, as well as potentials of the brain cortex (electroencephalography), are also biological rhythms. These intrinsic rhythms may themselves be subject to circadian or other extrinsic rhythms.

ADAPTATION, ACCLIMATIZATION (ACCLIMATION, ACCLIMATATION)

These terms characterize changes of form and function in biological organisms primarily as a response to environmental changes. The words are often used with little distinction, although definitions are available (46).

Because data are influenced in different ways when organisms react (adapt, acclimatize) to environmental changes, these effects are dealt with in a special section. Many viewpoints discussed here are also applicable to the section entitled Environmental Factors and vice versa.

Adaptation in physiology is mostly denoted as the response to environmental change within seconds or minutes (adaptation of sensory organs, such as dark adaptation of the retina of the eye). On the other hand, there is also the adaptation of evolution that takes place during millions of years; e.g., birds, fish, and mammals have developed a similar fish-like form (convergence) to reduce flow resistance when swimming (e.g. penguin, shark, dolphin).

Tropic and arctic plants and animals are "adapted" to hot or cold climates. The term "acclimatized" is correctly used in these cases. Llamas and chinchillas are adapted or acclimatized to high altitude.

On the other hand, short-term environmental changes, e.g., a sojourn in hot climates or high altitudes, cause reactions in individuals or species usually accustomed to cooler climates and lower altitudes. The increase of red blood cell count in mammals during high altitude sojourns is associated with "acclimatization." A man living near sea level exposed within an hour to a barometric pressure equivalent of 8800 m would die, but a man living for several months at 4000 m would survive this change. Natives of high altitude (Indians in Peru) would also survive. The lowlander "acclimatizes," whereas the native is already "acclimatized."

In mammals with sweat glands, acclimatization to hot climates leads to increased volume and decreased salt content of sweat. Plants and animals can adapt to salt concentration variations (fresh, brackish, sea-water) to a certain degree. Furthermore, cultivation of plants (wheat in colder climates) and breeding of animals (cow, chicken at high altitudes) in atypical environments are known.

Research data should indicate the state of adaptation or acclimatization of subjects or specimens if possible. Most data concerning these aspects are tabulated (35,47).

BIOELECTRICAL PHENOMENA

All living cells show membrane potential. The resting and action potentials of excitable tissues (nervous system, muscle) differ in size and duration from tissue to tissue. Some structures are "autorhythmic" such as the sinoatrial and atrioventricular nodes in the heart and many cells in the nervous system. The activity of cortex cells generates rhythmic potential changes (electroencephalogram). Many of these phenomena are tabulated (48,49), including electrocardiographic data (50-51).

MORPHOLOGICAL "DATA"

The documentation of morphological data is usually graphic and in two dimensions, although the subjects are three-dimensional. Figures are mainly measurements of length, height, diameter, thickness, surface, volume, or time of development (embryology, growth). A complete numerical description of a plant or an animal, though imaginable, would be extremely impractical. On the other hand, "scaling" in morphology serves many practical purposes (e.g., transplantation surgery). Therefore, scales should always be given in graphic documents whether dealing with a picture of a blue whale or the structure of a molecule.

Macroscopic Morphology

This covers biological material visible without optical instruments, documented in anatomical atlases of plants and animals. In embryology, visible morphological changes are correlated with developmental stages (33,52).

Microscopic Morphology

This covers biological material studies with light, phase, and electron microscopes; scales are usually provided. For embryology, such considerations as apply at the macroscopic level (e.g., somite counts) are also true here but at earlier stages of development (13,33,52).

Histochemistry and histoimmunology have introduced functional investigation into morphology. The effect of isolation of organs and cells (see section on the Influence of Experimental Procedures on Results), preparation of the animal, tissue sample, denaturation, shrinking by fixation, and changes of functions should be taken into account, if possible.

For practical reasons, morphological drawings often distort organs or parts of organs. This may lead inexperienced researchers in this field to wrong conclusions. For instance, the functional unit of the kidney, the nephron, is usually shown with loops of Henle that are much too short. This tempts the reader to underestimate the great total surface, and consequently the importance of the exchange capacity of this part of the nephron.

Submicroscopic Morphology

This is dependent only on indirect evidence to lead to the conclusion that at this level certain structures should be present or should have changed under certain conditions. "Receptors" in pharmacology and immunology are mostly receptors in a functional sense, but the site where they act can be estimated only indirectly (e.g., red blood cell surface and blood group systems).

Molecular Level Structural Analysis

These analyses (e.g., X-ray diffraction analyses) developed rapidly during the last decade. Composition and structure of enzymes, hormones, nucleic acids, proteins, and other complex biological molecules are well documented (53,54).

All restrictions (isolation of cells, preparation, analytical methods) mentioned above can play a role at the molecular level also. Generalizations can be misleading; for example, horse and fish hemoglobin structure or function are assumed to be the same.

METABOLIC PATHWAYS AND KINETICS

Metabolic pathways of plants and animals are qualitatively shown in rather large and (for the inexperienced) somewhat confusing maps (54-56). They represent what could be called the "anatomy" of metabolism.

Quantitative aspects cannot be derived from these maps. Concentrations and activities of metabolites, enzymes, etc., and kinetics of biological substances are partially tabulated (13,54,57).

The delicacy of these processes makes it necessary in the majority of investigations to work in vitro. These experiments often have to be performed at lower temperatures than were prevailing in the organism from which the organs, cells, and parts of cells were taken.

In warm-blooded animals especially, reaction rates (kinetics) are often too fast to be measured. Sometimes one has to investigate kinetics at temperatures as low as 2°C. However, methods for following fast reactions were developed, and have been refined during the last decade.

Research workers and reviewers must remember to mention in the original paper, and particularly in the tables, temperatures and any other "environmental" factors presumably deviating from the "normal" physiological situation.

PASSIVE EXCHANGE, ACTIVE TRANSPORT OF SUBSTANCES, PERMEABILITY

Exchange of substances in biological media, mostly through cell and other membranes (capillaries, tubules, etc.), can take place through several mechanisms: diffusion, facilitated diffusion, osmosis, active transport, pinocytosis. Often the data for transfer are expressed as fluxes of ions or electrolytes (58). Exchange ratios of substances are also listed under permeability of membranes, and some quantitative data are tabulated (13,57). Experimental conditions and source of material, of course, play an eminent role for these parameters.

PALEOBIOLOGY

A fascinating part of the biosciences, and of much greater importance than most biologists working with "living" material think, is the study of paleobiology. Comparative aspects of morphology and function in biology were, and still are, influenced by paleobiological research results. Ontogeny often is inspired by phylogenetic studies. Since dating in paleobiology is continuously subject to revision (59), the most recent dating, therefore, should be applied. Consult (60) for a survey of geologic time distribution of animals and plants with some references (61,62).

ETHOLOGICAL "DATA"

Ethology is a flowering branch of the biosciences with no exact relationship to psychology. Although most of the immense bulk of material can hardly be described by figures, there are many data which could be tabulated and hopefully will. Animal migration (e.g., birds) and homing (e.g., bees) are, more or less, well understood and can be partially expressed by figures or by a short description. The period of the imprinting process in ducklings can be estimated almost to the hour. A serious methodological difficulty faced by ethologists is that, as in other sciences, removing the animal from its natural environment changes its behavior and reactions. Also since most experiments are done in higher animals, the presence of the observer or apparatus may influence behavior. Telemetric methods cannot always exclude such influence, but television systems (also infrared) can reduce experimental influences.

COMPUTER APPLICATION TO BIOLOGICAL DATA COLLECTION

Increasingly, data are becoming computerized, especially in medicine, for diagnostic purposes: electroencephalography (63), electrocardiography (64,65), clinical chemistry (66), drug lists (67,68), toxic effects of chemical substances (67-69). In agriculture, crop genes and feeds are being registered (70). A computer compilation of taxonomic catalogs has been described (71), and a Laboratory Animal Data Bank has been established (72).

CLASSIFICATION, PROPAGATION, AND GEOGRAPHICAL DISTRIBUTION OF PLANTS AND ANIMALS, ABUNDANCE OF DIFFERENT SPECIES

In earlier sections, the importance of quoting the species or subspecies under investigation was stressed. The propagation of plants and animals is sometimes of importance to the investigator. The estimated number of living species is available (73).

The numbers of individuals in a class, a family, or a species is not as well documented, or even tabulated. They may be known only for man, animals, and plants menaced by extinction (74,75). The geographic distribution of plants and animals is documented (76-79). Such information can be helpful to someone interested in knowing how high above sea level a certain species of plant can be found in tropic zones (Andes) and in temperate zones (Alps), respectively. The same is true for animals, such as a family of rodents.

UNITS, GLOSSARIES, STATISTICS

It is self-evident that only well-defined units of measured data should be used (16). Unfortunately for many physical, physicochemical, and chemical data, several systems of units are in use. The coherent "Système International" (80) has not been officially accepted yet by governments of countries where scientists are generating vast amounts of data. Therefore, use of conversion formulas--even for such simple data as temperature (Fahrenheit to Celsius)--is necessary (81-87). For many units, symbols are being used; however, one symbol sometimes represents more than one unit (87).

Scientific terms should be well defined. Dictionaries and encyclopedias may help, but they prove less efficient as the terms become more specialized. Glossaries covering some fields of the biosciences have been published (88-96). If acceptable, one should follow such glossaries. Designations used to indicate the accuracy of data are published in the introduction to the *Biology Data Book* (1).

WHERE TO FIND UNTABULATED BIOLOGICAL DATA

When searching for data in the biosciences not yet compiled and tabulated, five categories of information sources should be consulted by the investigator:

- Handbooks in fields such as zoology, physiology, plant physiology, genetics (37,97-99). It can hardly be imagined that handbooks of even a comparatively small part of the biosciences (e.g., physiology) can continue to be published in the future. The fields are too broad, and rapid growth of scientific knowledge causes the first volumes to be, at least partly, out of date by the time the last volume is published. Nevertheless, taking into consideration the date of publication of a handbook article or table, as well as those publications located by consulting the following items below, it might still be useful to consult handbooks for data.

- *International Review of Science* (100) constitutes a new series with *International Review of Physiology*, (101), and *International Review of Biochemistry*, (102) representing the biosciences. It tries to cover all of science and is planning to report on progress in all fields covered every second or third year.

- A series of review serials, e.g., *Annual Review of Physiology* (103), publishes review articles on all aspects of a given field and covers, insofar as possible, all the recent references.

- Review literature in general. The multitude of other reviews published in a great number of periodic publications devoted entirely to reviews, as well as in scientific journals, is of a type different from the few discussed in the two preceding sections. The topics covered are chosen arbitrarily, for the most part, and are dependent on the editor's point of view concerning what is new and interesting. Thus, there may be topics which will not be covered for years to come. References to these reviews are accessible through on-line literature information services such as **MEDLARS, BIOSIS, SCISearch**, or their printed versions (104-107).

- Original articles and papers (proceedings) of symposia and congresses constitute the final source of published, but untabulated, data.

The sources listed above are best checked in sequence, keeping in mind that not every one will provide material for every research problem.

REFERENCES

1. Altman, P. L. and D. S. Dittmer, eds. *Biology Data Book.* 2nd ed. 3 vols. Federation of American Societies for Experimental Biology, Bethesda, Maryland (1972-1974).

2. Ibid. Vol. 1. pp. 490-495.

3. Bottle, R. T. and H. V. Wyatt. *The Use of Biological Literature.* 2nd ed. Archon Books, Hamden, Connecticut (1971).

4. In ref. 1, Vol. I. pp. 484-488.

5. Graham-Jones, O., ed. *Small Animal Anaesthesia.* Proceedings of a Symposium organized by the British Small Animals Veterinary Association. The Universities Federation for Animal Welfare, London, July 1963. Pergamon Press, Oxford (1964).

6. Jones, L. M. *Veterinary Pharmacology and Therapeutics.* 4th ed. Iowa State University Press, Ames, Iowa (1977). pp. 191-445.

7. Soma, L. R., ed. *Textbook of Veterinary Anesthesia.* Williams & Wilkins, Baltimore, Maryland (1971).

8. *De Haen Drug Interactions.* Vol. 1. Paul de Haen, Inc. New York (1972-) (Looseleaf).

9. Vasta, B. M., et al. ed. *Drug Interactions. An annotated bibliography with selected excerpts.* Vols. 1-3 (1967-1971). National Library of Medicine, Bethesda, Maryland (1972-1975) (DHEW Publication No. (NIH) 73-322, 75-322, and (FDA) 76-3004).

10. In ref. 1, Vol. III. pp. 1751-2040.

11. Dittmer, D. S., ed. *Blood and Other Body Fluids.* Compiled by P. L. Altman. Federation of American Societies for Experimental Biology, Bethesda, Maryland (1971).

12. In ref. 1, Vol. I. pp. 429-472.

13. Altman, P. L. and D. D. Katz *Cell Biology.* Federation of American Societies for Experimental Biology, Bethesda, Maryland (1976).

14. Ibid. pp. 61-73.

15. In ref. 1, Vol. I. pp. 517-527.

16. "Biologists' Guide for the Presentation of Numerical Data in the Primary Literature. Report of the CODATA Task Group on the Presentation of Biological Data in the Primary Literature." ***CODATA Bulletin*** No. **25** (November 1977).

17. Engler, A. *Syllabus der Pflanzenfamilien.* 12th ed. 2 Vols. Borntraeger, Berlin (1954-1964).

18. *Index Kewensis Plantarum Phanerogamarum, nomina et synonyma omnium generum...* 4 Vols. and suppls. Oxford University Press, Oxford (1893-1978).

19. Neave, S. A., ed. *Nomenclator Zoologicus: a list of the names of genera and subgenera from the 10th edition of Linnaeus 1758 to 1935.* 4 vols. and suppl. Zoological Society of London, London (1939-1976).

20. Rothschild, N. M. V. *A Classification of Living Animals.* 2nd ed. Longmans, London (1965).

21. Festing, M. F. W. and J. Staats, "Standardized Nomenclature for Inbred Strains of Rats." Fourth Listing. *Transplantation* **16**, 221-245 (1973).

22. Staats, J. "Standardized Nomenclature for Inbred Strains of Mice" Sixth Listing. *Cancer Res.* **36**, 4333-4377 (1976).

23. *Animals for Research. A directory of sources of laboratory animals, fluids, tissues, organs.* 8th ed. Institute of Laboratory Animal Resources, National Research Council, Washington, D.C. (1971).

24. Festing, M. F. W., Scientific ed. *International Index of Laboratory Animals.* 3rd ed. Laboratory Animal Centre, Medical Research Council, Carshalton, Surrey (1975).

25. Buchanan, R. E. and N. E. Gibbons, eds. *Bergey's Manual of Determinative Bacteriology.* 8th ed. Williams & Wilkins, Baltimore, Maryland (1974).

26. Fenner, F. *Classification and Nomenclature of Viruses;* 2nd report of the International Commission on Taxonomy of Viruses. Karger, Basel (1976). *Intervirology* **7**, 1-115 (1976).

27. Skerman, V. B. D. *A Guide to the Identification of the Genera of Bacteria.* 3rd. ed. Williams & Wilkins, Baltimore, Maryland (1975).

28. Martin, S. M. and V. B. D. Skerman, eds. *World Directory of Collections of Cultures of Micro-Organisms.* Wiley, New York (1972).

29. In ref. 3, pp. 358-361.

30. Mitruka, B. M. *Animals for Medical Research.* Wiley, New York (1976).

31. Birren, J., C. E. Finch and L. Hayflick, eds. *The Handbook of the Biology of Aging.* Van Nostrand, New York (1976).

32. Thorbecke, G. J. *Biology of Aging and Development.* Plenum Press, New York (1976). (FASEB monographs, 3).

33. Altman, P. L. and D. S. Dittmer, eds. *Growth, including Reproduction and Morphological Development.* Federation of American Societies for Experimental Biology, Bethesda, Maryland (1962).

34. In ref. 1, Vol. I. pp. 137-244.

35. Altman, P. L. and D. S. Dittmer, eds. *Environmental Biology.* Federation of American Societies for Experimental Biology, Bethesda, Maryland (1966).

36. In ref. 1, Vol. II. pp. 781-1052.

37. *Handbook of Physiology. A critical, comprehensive presentation of physiological knowledge and concepts.* American Physiological Society, Bethesda, Maryland (1959-1977). Sect. 4. "Adaptation to the environment." (1964). Sect. 9. "Reactions to environmental agents." (1977).

38. Robertshaw, D., ed. *Environmental Physiology. MTP International Review of Science: Physiology* Ser. I, Vol. 7. Butterworths, London (1974).

39. Spector, W. S., ed. *Handbook of Toxicology.* National Research Council, Committee on the Handbook of Biological Data. 5 Vols. Saunders, Philadelphia, Pennsylvania (1956-1959).

40. *Datensammlung zur Toxikologie der Herbizide. Hrsg. v.d. Arbeitsgruppe Toxikologie der Kommission für Pflanzenschutz-, Pflanzenbehandlungs - und Vorratsschutzmittel der Deutschen Forschungsgemeinschaft.* Verlag Chemie, Weinheim (1974). (Loose-leaf).

41. *Pesticide Handbook Entoma.* Entomological Society of America, College Park, Maryland (1977-1978). College Park, Maryland.

42. Henschler, D. ed. *Gesundheitsschädliche Arbeitsstoffe. Toxikologisch-arbeitsmedizinische Begründungen von MAK-Werten (Maximale Arbeitsplatzkonzentrationen).* Verlag Chemie, Weinheim (1972-). (Loose-leaf).

43. Roth, L. *Giftliste: Gifte, krebserzeugende, gesundheitsschädliche und reizende Stoffe.* Verlag Moderne Industrie, Munich (1976). (Loose-leaf).

44. In ref. 35, pp. 565-608.

45. In ref. 1, Vol. II. pp. 1015-1052.

46. Prosser, C. L., ed. *Comparative Animal Physiology.* 3rd ed. Saunders, Philadelphia, (1973).

47. In ref. 1, Vol. II, pp. 781-1052.

48. In ref. 1, Vol. II. pp. 1234-1260.

49. Remond, A., ed. *Handbook of Electoencephalography and Clinical Neurophysiology.* International Federation of Societies for Electroencephalography and Clinical Neurophysiology. Vols. 1-16. Elsevier, Amsterdam (1971-).

50. Altman, P. L. and D. S. Dittmer, eds. *Respiration and Circulation.* Federation of American Societies for Experimental Biology, Bethesda, Maryland (1971). pp. 270-299.

51. In ref. 1, Vol. III, pp. 1730-1738.

52. In ref. 1, Vol. I., pp. 173-222.

53. Dayhoff, M. O. and R. V. Eck, ed. *Atlas of Protein Sequence and Structure*, Vol. 5, suppl. 1-3. National Biomedical Research Foundation, Silver Spring, Maryland (1972-1977).

54. Fasman, G. D., ed. *Handbook of Biochemistry and Molecular Biology: Physical and Chemical Data*. 3rd ed. 2 Vols. Chemical Rubber Company Press, Cleveland, Ohio (1976).

55. In ref. 1, Vol. III. pp. 1530-1569.

56. Michal, G. *Biochemical Pathways*, Rev. ed., Boehringer, Mannheim, Federal Republic of Germany (1974).

57. Altman, P. L. and D. S. Dittmer, *Metabolism*. Federation of American Societies for Experimental Biology, Bethesda, Maryland (1968).

58. In ref. 1. Vol. II. pp. 1236-1237, 1241-1242.

59. Cain, W. F. and H. E. Suess, "Carbon 14 in Tree Rings" *J. Geophys. Res.* **81**, 3688-3694 (1976).

60. In ref. 1, Vol. I. pp. 533-534.

61. Boureau, E., ed. *Traité de Paléobotanique*. Vol. 2-4,2. Masson, Paris (1964-1975).

62. Piveteau, J., ed. *Traité de Paléontologie*. 7 vols. Masson, Paris (1952-1969).

63. Dolce, G. and H. Kuenkel, *CEAN Computerized EEG Analysis*. Symposium of Merck'sche Gesellschaft für Kunst und Wissenschaft, Kronberg/Taunus, April 8-10, 1974. Fischer, Stuttgart (1975).

64. Van Bemmel, J. H. and J. L. Willems, eds. *Trends in Computer-Processed Electrocardiograms*. Proceedings of the IFIP-TC-4 Working Conference, Amsterdam, 3-5 November, 1976. North-Holland, Amsterdam (1977).

65. Zywietz, C. and B. Schneider, eds. *Computer Application on ECG and VCG Analysis*. Proceedings of the 2nd IFIP-TC-4 Working Conference on Computer Application on ECG and VCG Analysis, Hannover, 11-14 Oct. 1971. North-Holland, Amsterdam (1973).

66. Heu, R., J. Otten, and H. D. Purps, "Ein autarkes Datenverarbeitungs-Subsystem für das klinisch-chemische Laboratorium." *Roentgenstrahlen* **30**, 25-31 (1974).

67. Windholz, M., ed. *Merck Index. An Encyclopedia of Chemicals and Drugs*. 9th ed. Merck & Co., Rahway, New Jersey (1976).

68. *Rote Liste. Verzeichnis pharmazeutischer Spezialpräparate*. Ed.: Bundesverband der Pharmazeutischen Industrie, Frankfurt, Federal Republic of Germany (1976).

69. National Institute for Occupational Safety and Health (NIOSH). *Registry of Toxic Effects of Chemical Substances (RTECS)*. 7th ed. U.S. Government Printing Office, Washington, D.C. (1977). Available also: On-line data retrieval file via National Library of Medicine, Bethesda, Maryland

70. Tomberg, A., ed. *Data Bases in Europe. A directory to machine-readable data bases and data banks in Europe*. 3rd ed. Association of Special Libraries and Information Bureaux (ASLIB), London (1977). p. 27.

71. Raworth, D. A. and B. D. Frazer, "Compilation of Taxonomic Catalogues by Computer." *J. Entomol. Soc. B. C.* **73**, 63-67 (1976).

72. *Laboratory Animal Data Bank (LADB).* National Library of Medicine, Bethesda, Maryland.

73. In ref. 1, Vol. I. pp. 528-529.

74. *International Union for Conservation of Nature and Natural Resources. Survival Service Commission. Red Data Book.* Vols. 1-5. Morges, Switzerland (1966 -). (Loose-leaf).

75. Melby, E. C., Jr. and N. H. Altman, *Handbook of Laboratory Animal Science.* Vol. 2. pp. 462-476. Chemical Rubber Company Press, Cleveland, Ohio (1974).

76. Andrewartha, H. G. and L. C. Birch *The Distribution and Abundance of Animals.* University Press, Chicago (1954).

77. Cox, C. B., I. N. Healey, and P. D. Moore, *Biogeography: an ecological and evolutionary approach.* 2nd ed. Wiley, New York (1976).

78. Takhtadzhian, A. L. *Flowering Plants: Origin and dispersal.* Transl. from the Russian. Smithsonian Institution Press, Washington, D.C. (1969).

79. Udvardy, M. D. F. *Dynamic Zoogeography, with special reference to land animals.* Van Nostrand, New York (1969).

80. *Le Système International d'Unités (SI),* 3rd ed. Bureau International des Poids et Mesures, Sèvres, France (1977). Engl. translations: a) London. Her Majesty's Stationery Office. b) International System of Units (SI). Washington, D.C., U.S. Government Printing Office, August 1977. (NBS Special Publication, No. 330). Stock No. 003-003-01784-1.

81. Altman, P. L. and D. S. Dittmer, eds. *Biology Data Book.* Federation of American Societies for Experimental Biology, Bethesda, Maryland (1964). pp. 571-578.

82. In ref. 1, Vol. I, pp. 534-538.

83. Diem, K. and C. Lentner, eds. *Documenta Geigy. Wissenschaftliche Tabellen.* Vol. 1. 7th ed. Thieme, Stuttgart (1975). Also 8th ed., rev. and enlarged. Ciba-Geigy, Basel (1977).

84. *International Commission on Radiation Units and Measurements: Dose Equivalent.* Suppl. to ICRU Report 19. Washington, D.C. (1973).

85. "Interunion Commission on Biothermodynamics. Recommendations for Measurement and Presentation of Biochemical Equilibrium Data." *J. Biol. Chem.* **251,** 6879-6885 (1976). (also: ***CODATA Bulletin*** No. **20** (Sept. 1976).)

86. Lippert, H. and H. P. Lehmann, *SI Units in Medicine.* Urban & Schwarzenberg, Baltimore, Maryland (1977).

87. Lowe, D. A. *A Guide to International Recommendations on Names and Symbols for Quantities and on Units of Measurement.* WHO, Geneva (Bulletin of the WHO, 52nd Suppl. "Progress in standardization," 2 (1975)).

88. Bartels, H., P. Dejours, R. H. Kellogg, and J. Mead, "Glossary on Respiration and Gas Exchange." *J. Appl. Physiol.* **34,** 549-558 (1973).

89. Bligh, J. and K. G. Johnson "Glossary of Terms for Thermal Physiology," *J. Appl. Physiol.* **35**, 941-961 (1973).

90. Faulkner, W. F., J. W. King, and H. C. Damm, eds. *CRC Handbook of Clinical Laboratory Data.* 2nd ed. Chemical Rubber Company Press, Cleveland, Ohio (1968). pp. 591-608. "Glossary of Microbiological Terms."

91. Frear, D., ed. *Pesticide Index: A glossary of pesticide chemical names.* 4th ed. College Science Publications, State College, Pennsylvania (1969).

92. Law, J. W. and H. J. Oliver *Glossary of Histopathological Terms.* Butterworths, London (1973).

93. Rieger, R., A. Michaelis, and M. M. Green *Glossary of Genetics and Cytogenetics. Classical and molecular.* 4th ed. Springer, Berlin (1976).

94. Samson, P. *Glossary of Haematological and Serological Terms.* Appleton-Century-Crofts, New York (1972).

95. Samson, P. *Glossary of Bacteriological Terms.* Butterworths, London (1975).

96. Van der Hammen, L. *Glossaire de la Terminologie Acarologique. Glossary of Acarological Terminology.* 2 Vols. Junk, The Hague (1976).

97. Pirson, A. and M. H. Zimmermann, eds. *Encyclopedia of Plant Physiology.* New Series. Springer, Berlin (1975-).

98. Grassé, P. P., ed. *Traité de Zoologie: Anatomie-systematique-biologie.* 17 Vols. Masson, Paris (1948-1975).

99. King, R. C., ed. *Handbook of Genetics,* 5 Vols. Plenum Press, New York (1974-1976).

100. *International Review of Science* (A series of frequently updated reviews). MTP Press, Lancaster; Butterworths, London; University Park Press, Baltimore.

101. Guyton, A. C. ser. ed. *International Review of Physiology.* Series II, Vols. 9-16. University Park Press, Baltimore, Maryland; MTP Press Ltd., Lancaster, England (1976-1978). 9. "Cardiovascular physiology II"; 10. "Neurophysiology II"; 11. "Kidney and urinary tract physiology II"; 12. "Gastrointestinal physiology II"; 13. "Reproductive physiology II"; 14. "Respiratory physiology II"; 15. "Environmental physiology II"; 16. "Endocrine physiology II."

102. Kornberg, H. L. and D. C. Phillips, ser. ed. *International Review of Biochemistry.* Series II, Vols. 13-25. University Park Press, Baltimore, Maryland; MTP Press Ltd., Lancaster, England (1977-). 13. "Plant Biochemistry II"; 14. "Biochemistry of Lipids II; Chemistry of macromolecules II; Physiological and pharmacological biochemistry II"; 16. "Biochemistry of nucleic acids II; Biochemistry of cell differentiation II"; 17. "Amino acid and protein biosynthesis II; Biochemistry of carbohydrates II"; 20. "Biochemistry and mode of hormones II; Defense and Recognition II; Biochemistry of nutrition I; Biochemistry of cell walls and membranes II; Microbial biochemistry I."

103. *Annual Review of Physiology.* Vol. 1- Annual Reviews, Palo Alto, California (1939-). (For list of other fields of science covered annually by this publisher see Chapter 2, Bibliography B-1.

104. *Bibliography of Medical Reviews.* National Library of Medicine, Bethesda, Maryland (Included as a separate section in each monthly *Index Medicus.*)

105. *Biological Abstracts.* Biosciences Information Service of Biological Abstracts, Philadelphia, Pennsylvania.

106. *Bioresearch Index.* Biosciences Information Service of Biological Abstracts, Philadelphia, Pennsylvania.

107. *Index to Scientific Reviews. An interdisciplinary index to the review literature of science, medicine, agriculture, technology, and the behavioral sciences.* Institute for Scientific Information, Philadelphia (1974-).

Data Handling for Science and Technology
S.A. Rossmassler and D.G. Watson (eds.)
North-Holland Publishing Company

TREATMENT OF OBSERVATIONAL DATA IN THE GEOSCIENCES: ESTIMATION AND APPROXIMATION OF DATA

A. H. Shapley and R. Tomlinson

ABSTRACT

Geosciences data are acquired, handled, and applied in very large quantities; the volume of data involved makes for qualitative differences in the techniques employed. Observational data are described in three categories: monitoring data, survey data, and complex observations. Storage and dissemination may be preceded or paralleled by "massaging" which is designed to facilitate storage or subsequent analysis. Analysis of such data involves both straightforward statistical treatment and more advanced approaches when the data sets are multidimensional.

INTRODUCTION

In this book, as in many other multidisciplinary communications on the subject of "data," the reader will notice that the word "data" is being used with somewhat different shades of meaning in different parts of the text. Workers in other disciplines, of which perhaps physics and chemistry are core examples, sometimes use the word to refer to fundamental values, the unvarying facts on which the laws of their disciplines rest; in the geosciences the raw data often consist of time- and space-dependent records of observations and experiments. Such data must subsequently be analyzed, reduced, and linked to an observational framework before they can be interpreted. Accordingly, in the astro- and geosciences, the results of any quantitative (and even qualitative) observations are regarded as the "data" upon which the investigators of the discipline rely. This state of affairs is a natural result of the different objectives and lines of investigation pursued in different disciplines but may tend to confuse the reader. To put the following discussion on data handling in the geosciences into perspective, the term "data" should be interpreted in its fuller sense. "Data" as referred to in this chapter may be defined as the numerical and/or graphical results of any of the wide variety of observations and data reduction techniques employed in the geosciences.

The problems of handling observational data such as those obtained in the geosciences are, as the preceding paragraph suggests, quite different from those encountered in laboratory physics and chemistry. Accordingly, they are given separate treatment in this ***Sourcebook***. For example, there is a huge quantity of data involved: one satellite image of terrestrial cloud cover contains 10^8 data points; and images are taken every few minutes--mounting up to 10^{12} bits per year per satellite. Monitoring the earth's magnetic field, the ionosphere, or earth motions by seismographs by some 200-600 stations throughout the world accumulates only slightly less data. Modern seismic soundings of the ocean bottom lead to the accumulation of data at the rate of some 10^4 per kilometer, and there are over 10^6 cruise kilometers made each year. Thus it is readily apparent that, if a large part of these data is considered at one time, the largest computers in the world must be used for modeling the dynamics of the geosciences areas. In addition, the largest mass storage capacity must also be available. The techniques for handling these kinds of space-varying and time-varying data are quite different from those for most kinds of laboratory data. However, many of the basic principles are the same for observational and laboratory data handling, and some of the techniques can apply to either case. We will try to cover some of these aspects in this chapter.

TYPES OF OBSERVATIONAL DATA

Observations in the geosciences fall into three categories:

- *Monitoring*, or repeated observations of the same parameters at a given location under local or global conditions which change with time;

- *Surveys*, or repeated observations of the same parameter successively at different locations under conditions which do not change with time, or the time changes are slow compared to the duration of the observational program;

- *Complex geoscience observations*, in which there is a mixture of time and space variations.

Monitoring data come from the fixed observatories which measure atmospheric parameters, the short-time variations of the earth's magnetic field, volcanic activity, and the like. The time interval between observations can range from fractions of a second to hours, days, or longer. The variations of interest may be essentially continuous or sometimes episodic. The time resolution of scientific or practical significance may be commensurate with the potential measuring capability, or may be much less. For example, 10-second interval data on temperature are not necessary or useful for the local weather station, while they are essential for monitoring geomagnetic variations during a sunstorm. Thus both the design of the "observational experiment" and the handling of the resulting data depend on the scientific or practical problem under study.

A typical "one-dimensional" survey would be an ocean cruise during which repeated measurements would be made of gravity or water temperature at a specified depth along the path of travel. Some cruise data are "two-dimensional," as in the case of side-scan sonar observation of the ocean bottom topography. More common examples of surface or "two-dimensional" data are aerial photographs and satellite images. Of course, the data from original "one-dimensional" surveys are frequently combined through various data handling techniques into two-dimensional surveys. Three-dimensional surveys include solid earth studies and, on climatological time scales, synthesized studies of the atmosphere.

Many geoscience phenomena vary in both time and space. In some cases the observational data are mixed, such as in situ atmospheric variability measurements from a low altitude satellite. In most cases, however, the physics of the phenomena must be pieced together through data handling by the consideration of repeated surveys, as in aerial or satellite data on forestation, or through combining monitoring data from many locations, as in ground-based meteorology.

The characteristic that distinguishes many geoscience data is that they are not repeatable under identical or completely controlled conditions. Also, as a practical matter, it is not desirable to evaluate each observation individually; rather more general approaches are taken in estimating precision, accuracy, and errors.

DATA STORAGE AND DISSEMINATION

A distinguishing feature of observational data is the variety of data forms used. While many data are digitized to begin with, many others are recorded and even used directly and stored in analog form -- strip charts or photographs. The latter are often scaled or otherwise converted to digital or sometimes descriptive form, after selection and interpretation through human judgment. Still another feature is that the scientific literature is not the principal medium for disseminating many large bodies of geoscience data; the quantities are too great, and the needs for dissemination can, in such cases, be better met through the use of alternate (e.g., computer-compatible) formats. Accordingly some categories of geoscience data are disseminated primarily through scientist-to-scientist exchange or through data centers (1).

Usually the data disseminated are selections from the total file. Most seismologists study seismograms only for earthquakes of special interest to them individually, and do not want seismograms for days without earthquakes. Thus there are problems of selective retrieval from a large homogeneous data set.

Much of the geoscience data is stored in analog form, even if it were originally digital. Analog records are often the more compact and more easily used format for their anticipated uses. Machine-readable data are commonly stored serially on tape, although permanent storage on disk is more efficient for users who treat large time- and space-varying data files. Multidisciplinary data stores are becoming common in addition to single-parameter files.

DATA MASSAGING

Data massaging techniques are used to change the form of data so that they may be more readily stored or may be more amenable to subsequent analysis. Included are: smoothing of lines and boundaries, the elimination of distortion in graphic records, the rotation and translation of coordinate systems, the matching of edges between adjoining sheets of graphic data, and the rectification of topological errors.

Data smoothing is the process of moderating changes in the intensity of data values at a certain point, through the application of a factor based on the relative intensity of data values at points in the surrounding neighborhood.

Data smoothing implies a potential reduction in the information content of the data to achieve the benefit of a new data set with fewer spurious values, potentially reduced volume, improved data compactness, and subsequent ease of processing. In any technique which may degrade the information content of raw observational data, it is important to weight the potential loss involved against the intended use of the data, potential reexpansion and desired calculation or storage accuracy, before deciding to employ the technique concerned.

One-dimensional smoothing is the removal of points in a sequence of data points along a line. It can be thought of by reference to an imaginary line traced across a regular matrix of points. "Linear single point smoothing" is the removal of the intermediate point whenever point n + 2 is within one grid square of point n on the line. The following diagram illustrates the results.

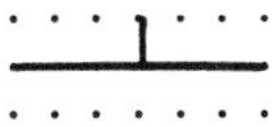

Spike becomes

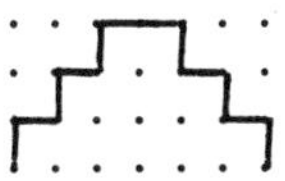

Zig-zag becomes

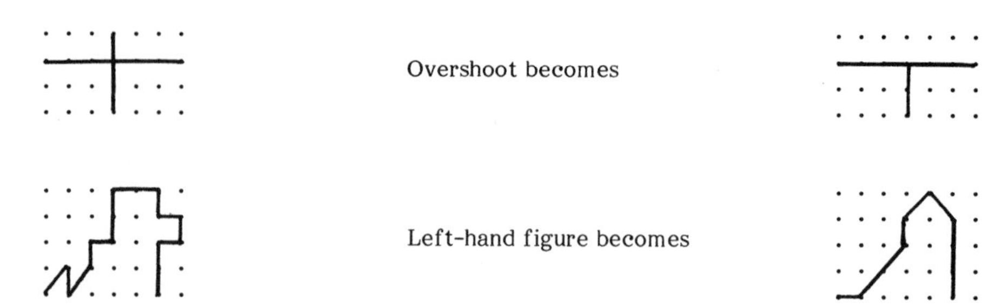

"Line following" is a related process that can be envisaged as selecting "important" points along a line so that the shape of the line is preserved while reducing the number of points needed to describe it. Montanari (2) described a straightforward approach; this has since been improved upon by Kingston (3). Essentially, each point on a line is assigned an imaginary value which is a measure of its adjacency and contiguity with neighboring points. Points with the highest values are then entered and the rest are discarded.

Just as spurious "spikes" and "overshoots" are removed by data smoothing, so can small gaps in lines (sequences) and surfaces (arrays) be automatically closed.

A sophisticated extension of data smoothing and the re-evaluation of data points with respect to neighboring data result in the new field of image enhancement and picture processing (4). It is emphasized that these techniques are not in themselves data analysis but rather a change in the data to facilitate subsequent analysis.

"Normalization" of data is another approach aimed at reducing extremes of data that are not meaningful. Usually normalization means subtracting the mean of a data set from each data point in the set and dividing the result by the standard deviation.

"Distortion" of stretched images, such as that caused by the shrinkage in a paper record of data, can be eliminated or measured by "rubber sheet" stretching (5). A simple illustrative example involves the reconstitution of digital data derived from scanning a map with known borders. These borders normally are rectangular, or with straight sides and curves for top and bottom edges (as, for example, on maps based on a Universal Transverse Mercator projection). By fitting the sides, top and bottom to known formulae, a transformation coefficient matrix may be generated. This matrix may be applied as a mapping transformation to adjust the whole image fairly quickly.

Similarly, a set of data descriptive of a surface may need to be "rotated" with respect to a cardinal direction, or to be transferred from one map projection to another so that it may be fitted over another set of surface data for storage or comparison. Rotation is achieved by calculating the rotation angle of each boundary "side" and moving the points within the map according to their related angular displacement. Such rotations may be done on a point-by-point basis, but because this can be laborious even with modern computer technology, the process is often accomplished small-section-by-small-section of the map. A colinear transformation (so named because any straight line remains a straight line after undergoing the transformation) is applied within each section. The resulting errors are small if the sections are small.

"Edge matching" is needed because arrays of data frequently have to be broken into sections due to limitations of computer memory size. Various techniques are employed, many based on assigning special codes to the truncated ends of curves that cross the sector border and an iterated matching of those codes during subsequent processing. The task is mainly of importance when data are in a vector format. The current exploration (6) of raster data manipulation maintains that edge-matching is a trivial process when data sets are matrix rather than lineal.

"Data rectification" is often a necessary process to apply to a data set to eliminate errors introduced by the instruments used to sense and record the data. Perhaps the best known example is the rectification of aerial photographic images that have been recorded while the aircraft platform was tilted. Until recently, the approach used to remove the distortion was to position mechanically the transparency of the photograph so that its new projected image fitted over the position of known ground control points plotted on the projection screen. The process of rectifying digital images is conceptually the same. Certain points in the data set are identified with known positions on the ground and the image is moved digital-point-by-digital-point until all points are placed in accordance with the known points.

A sophisticated extension of these data massaging techniques allows data from observations taken from constantly moving satellite platforms to be provided in a form that will overlay a standard topographic map for ease of subsequent analysis.

It is emphasized that the examples of data massaging provided above are not exhaustive. They do, however, illustrate the techniques that are frequently employed to assist the worker in subsequent data analysis.

Two broader types of data handling problem stem from the nature of the raw data generated by observational techniques in the geosciences. The first of these arises from the fact that much of the basic data is in graphic form, or has traditionally been converted to a graphic form even after original measurements were available in numerical form. The term "graphic form" here encompasses the wide variety of record formats, including paper tapes from instruments, photographic images, satellite images, maps and charts, that are used to hold much of geoscience data. Modern computers demand that data be in machine-readable numerical form and the problem of transforming graphic data to an acceptably error-free, useful, machine-readable form (a process frequently referred to as digitizing) is still one fraught with difficulty and expense. Discussions of the techniques involved in encoding the data in numerical form can be found in several works (7,8). The various options for the physical transformation of data formats are discussed in recent publications (9,10).

The second problem is that the volume of data now in digital form is immense and data gathering institutions are committed to further growth. The U.S. Geological Survey, for example, has over 50 systems handling a wide variety of geoscience data in digital form. The aggregate volume of such data already in machine-readable form in 1977 was approximately 500,000 million bits. (A bit is the smallest unit of normal machine-readable information. One regular reel of magnetic tape can hold up to 700 million bits.) Conservative estimates indicate that this will grow by more than 250 per cent to 1.7 million million bits by 1981. Other institutions have similar objectives and growth patterns. It can be assumed that computers will become better and cheaper, and that the developing processes of institutional management will tend to match data production to handling capability, or more particularly to computing capacity. There is, however, cost associated with the use of computers, and the volume of data to be processed has a marked impact on that cost. We are rapidly passing the point where it can be assumed that "a computer will handle it" and questions are being asked of how to organize large volumes of geoscience data for efficient subsequent examination. Discussion of these problems and the role of data base management systems can be found in recent work (10-12).

DATA ANALYSIS TECHNIQUES

(A general overview of this topic is provided in Chapter 4 "Analysis and Interpretation of Data.")

Geoscience suffers from the lack of an adequate body of general theory about the nature of the populations of entities that it examines. The majority of analysis techniques are in fact based on the premise that information about a phenomenon can be deduced from an examination of a small sample collected from a vastly larger set of potential observations of the phenomenon.

Many of the observations of geoscience result in data values that are space and time dependent and the type of data analysis employed is related to the dimensionality of the data. Furthermore, many geoscience observations are made at sites removed from the bulk of their study material, and workers must rely on data in which there are substantial amounts of uncertainty. If the data are statistically independent, a wide variety of standard elementary data analysis techniques can be applied, to start to understand better the nature of the data, the level of uncertainty, and the character of the population being examined. Typical procedures include probability analysis, the use of statistics to estimate the parameters of parent populations and to test hypotheses about populations, the analysis of distributions, and analysis of variance. These are straightforward procedures that are well documented in the introductory reference literature (13-18) and need no further explanation here. Perhaps the main caution in their use in the geosciences is that one be certain that the sometimes rigorous assumptions underlying the different procedures are met before the results of the analysis are relied upon. This is not always the case in the literature.

Data sets with one or more dimensions pose a somewhat greater analytical challenge (19,20). One-dimensional data in space or in time can perhaps be visualized as a series of data values occurring in sequence along a line. Two-dimensional data can be thought of as a set of data values distributed on a surface (typically involving the analysis of data derived from maps, charts, and photographs). Much of the same risk of uncertainty about the samples encountered in spatial data has to be tackled in this new type of data. Similarly, there is an interest in the general tendencies exhibited by the one- and two-dimensional data so that one can interpolate between data points, extrapolate beyond the data sequence, infer the presence of trends, or estimate the characteristics that may be of interest to the earth scientists.

Interpolation procedures for data along a line are fairly straightforward in concept. The two-dimensional version is, of course, the process involved in contouring a set of points (21-25). The first step in interpolation is usually producing a regularly spaced set of values. When contouring a surface, values at regular grid points may be produced in a variety of ways, ranging from estimates derived from the nearest originally observed values, to the estimates derived from the fitting of a trend surface. The former approach results in a rather coarsely approximated set of contours, the latter has the disadvantage that none of the original points are likely to be on the generated contours.

Frequently one is not only interested in the magnitude of changes in a sequence or on a surface, but also curious about where these changes occur. If certain assumptions can be justifiably made about the distribution of the population on which the samples are collected, statistical tests called *"regression analysis"* can be performed (26). The simplest of these is the family of least-squares methods for determining a line about which the variance is at a minimum. Linear regression techniques are applied to fit a line through a series of points with a controlled, or at least a well-understood, degree of fit.

The process of *linear data smoothing* was mentioned in the section on "data massaging" above. It is easy to imagine a process of moving averages of data values along a line that results in a smoother line. Somewhat more revealing is the analysis that results in separating the underlying (long-term) variations in values along a line from local (short-term) perturbations. This latter process is known as *"filtering"* (27) or time-trend analysis. The two-dimensional version of this process is *trend-surface analysis,* (19,28-30) which is most frequently applied to separate the general or regional fluctuations in curvature of a surface from local crinkles in the surface.

An extension of this two-dimensional surface analysis involves taking into account a third independent geographic coordinate (and all its effects on the other coordinates). The resulting regression is sometimes referred to as a hypersurface or a "four-dimensional trend surface." When the original surface is defined by irregularly-spaced sample points, the solution is computationally more involved (31). This line of analysis is increasingly dependent on computer-based solutions, and there are available several programs which will compute and plot trend surfaces up to even the sixth degree (32).

A pattern in variation of data values along a line will sometimes appear to repeat itself and one may become curious about the repetitive nature of the cycle. Two similar parts of a sequence of values can be conceptually overlaid and the goodness of fit examined. This is in essence the process of *"auto-correlation,"* (33) except that in the latter, all parts of the sequence are compared to itself at all possible positions and a coefficient of correlation is calculated. This gives an excellent appreciation not only of goodness of fit but also badness of fit. The process of *cross-correlation* is a similar operation, but between two different sequences of data.

A variation of this examination of the cyclical properties in a sequence of data is the recognition of the dominant periodic components. This is termed *Fourier analysis* (12,27,29,34). Again, there is a two-dimensional version of the process. Just as the single Fourier process looks for the periodic variations in a linear sequence, so does double Fourier series analysis look for the wavelengths of ripples in a surface.

The concepts of moving averages in a sequence can, of course, be applied to a surface (35) One particularly useful approach, which was developed in mining geology and is now utilized in many earth sciences, is termed *"Kriging"* after the South African mining geologist D. G. Krige (36) This technique not only attempts to estimate the values of a spatially distributed variable but also to assess the probable error associated with the estimates.

Comparison of surfaces (comparison of maps, for example) can be as simple as determining the difference in elevation of two sets of contours for one region. The result is a third map showing the similarity of the two surfaces. The task gets harder when the two surfaces contain different types of data (lines on surfaces, points on polygons) which may be expressed in different units and corrected from different control points. Essentially, one surface has to be transferred to the terms of the other and the comparison takes place after the maps have been standardized. One can similarly compare two trend surfaces, two hypersurfaces, and so on.

The problem of adding more dimensions to the data set brings us into the realm of *"multivariate analysis"* (37,38,39). Here the object is to compare the changes in several properties simultaneously. Some applications are extensions of the methods described above; other approaches are totally new. *Multiple regression* (26) represents an extension to multivariate data. The calculation of *discriminant functions* consists of finding a transform which gives the minimum ratio of the difference between a pair of group multivariate means to the multivariate variance within the two groups. *Cluster analysis*

(40) is simply the clustering of multivariate entities into more or less homogeneous groups. *Factor analysis* (41,42), and the closely-related technique of *principal components analysis*, is concerned with examining and describing the structure within a set of multivariate data. Essentially, the approach is an analysis of correlations. It extracts vectors from the matrix of correlations and covariances, and the percentage of the total variability explained by these vectors is an indication of the interrelationships within the data set.

REFERENCES

1. See for example, Appendix I, World Data Center System.

2. Montanari, V. "A Note on Minimal Length of Polygonal Approximation to a Digitized Contour." *Comm. of A. C. M.*, **13**, 41 (1970).

3. Kingston, P. P. C. "Data Massaging." *Geographical Data Handling*, R. F. Tomlinson, ed., International Geographical Union Commission on Geographical Data Sensing and Processing, Ottawa, Canada (1972) p. 740.

4. Rosenfeld, A. *Picture Processing by Computer.* Academic Press, New York (1969).

5. Linstedt, C. *Preliminary Experiments on Geometric Transformations of Images.* IBM Res. RC 3314, IBM Research Center, Yorktown Heights, New York (1971).

6. Peuquet, D. J. "Raster Data Handling in Geographic Information Systems." *Proceedings, Symposium on Topological Data Structures for Geographic Information Systems,* Harvard University, Cambridge, Massachusetts (1977).

7. Dueker, K. J. "A Framework for Encoding Spatial Data." *Geographic Analysis* **4**, 98-105 (1972).

8. Freeman, H. "On the Encoding of Arbitrary Geometric Configurations." Institute of Radio Engineers, *Transactions on Electronic Computers,* **EC-10**, 260-268 (1961). (Seminar paper on digital encoding of spatial data.)

9. Tomlinson, R. F., ed. *"Geographical Data Handling."* International Geographical Union Commission on Geographical Data Sensing and Processing, Ottawa, Canada (1972) 1300 pp. (Paperback source book on geographical data storage, handling, analysis, and display.)

10. Tomlinson, R. F., H. W. Calkins, and D. F. Marble, "Computer Handling of Geographical Data." *Natural Resources Research Series XIII,* Unesco, Paris. (1976).

11. Dodd, G. G. "Elements of Data Management Systems." *Computing Surveys* **1**, 117-133 (1969).

12. Martin, J. *Principles of Data Base Management.* Prentice-Hall, Englewood Cliffs, New Jersey (1976) 352 pp. (An excellent, well-illustrated introductory text.)

13. Wallis, W. A. and H. V. Roberts, *Statistics, A New Approach.* The Free Press, Glencoe, Illinois (1956) 646 pp. (Statistics without tears.)

14. Davis, J. C. *Statistics and Data Analysis in Geology.* John Wiley, New York (1973) (Excellent, clearly written introduction to topic, well selected references.)

15. Panofsky, H. A. and G. W. Briev, *Some Applications of Statistics to Meteorology.* Pennsylvania State University, University Park, Pennsylvania (1965) 224 pp.

16. Duckworth, W. E. *Statistical Techniques in Technological Research.* Methuen,

London (1968). (Excellent introduction to basic statistical procedures.)

17. Freund, J. E. and F. J. Williams, *Dictionary Outline of Basic Statistics.* Oliver and Boyd, Edinburgh (1966) 175 pp. (Good paperback.)

18. Koch, G. S., Jr. and R. F. Link, *Statistical Analysis of Geological Data*, Vol. 1. John Wiley, New York (1970) 375 pp. (Univariate statistical analysis.)

19. King, L. J. *Statistical Analysis in Geography.* Prentice-Hall, Englewood Cliffs, New Jersey (1969) 288 pp. (Advanced level spatial analysis.)

20. Tobler, W. R. "Numerical Map Generalizations and Notes on the Analysis of Geographical Distribution." *Discussion Paper 8, Michigan Inter-University Community of Mathematical Geographers,* University of Michigan, Ann Arbor, Michigan (1966).

21. Holroyd, M. T. and B. K. Bhattachayya, "Automatic Contouring of Geophysical Data Using Bi-cubic Spline Interpolation." Paper 70-55, Geological Survey of Canada, Ottawa (1970).

22. Business Machines, Data Processing Applications, White Plains, New York (1965). (Beginners book on contouring.)

23. Murray, F. W. *A Method of Objective Contour Construction.* The RAND Corporation NTIS No. AD 666,830 (1968).

24. Shepard, D. "A Two-Dimensional Interpolation Function for Computer Mapping of Irregularly Spaced Data." *Harvard Papers on Theoretical Geography,* No. **15**, (1968).

25. Walters, R. F. "Contouring by Machine: A Users Guide." *Amer. Assoc. of Petroleum Geologists, Bull. V.* **53**, 2324-2340 (1969).

26. Draper, N. R. and H. Smith, *Applied Regression Analysis.* John Wiley, New York (1966). 407 pp. (Excellent text on multiple regression analysis.)

27. Robinson, J. E. "Spatial Filtering of Geological Data." *Review of the Technical Statistical Inst.*, **38**, 21-32 (1970). (Fourier analysis and spatial filtering of contour maps.)

28. Chorley, R. J. and P. Haggett, "Trend Surface Mapping in Geographical Research." *Trans. Inst. British Geographers,* No. **37**, 47-67 (1965).

29. Harbaugh, J. N. and D. F. Merriam, *Computer Simulation in Stratigraphic Analysis.* John Wiley, New York (1968) 202 pp. (Trend surface analysis and Fourier analysis.)

30. Merriam, D. F. and N. C. Cockes, eds. *Computer Applications in the Earth Sciences: Colloquium on Trend Analysis.* Kansas Geological Survey Computer Contribution 12, (1967) 62 pp.

31. Krumbein, W. C. "Trend Surface Analysis of Contour-Type Maps with Irregular Control-Point Spacing." *J. Geophys. Research,* **64(7)** (1959).

32. O'Leary, M., R. H. Lippert, and O. T. Spitz, *FORTRAN IV and MAP Program for Computation and Plotting of Trend Surfaces for Degrees 1 through 6.* Computer Contribution 3, State Geological Survey, University of Kansas, Lawrence, Kansas (1966).

33. Cliff, A. D. and J. K. Ord, *Spatial Autocorrelation.* Pion, London (1973). (The most comprehensive text on the subject to date.)

34. Davis, J. C. and M. J. McCullagh, *Display and Analysis of Spatial Data.* John Wiley, New York (1975). 378 pp. (Collection of papers, well chosen and now hard to find individually.)

35. Matheron, G. "Principles of Geostatistics." *Econ. Geology,* **58**, 1246-1266 (1963).

36. Krige, D. G. "Two-dimensional Weighted Moving Average Trend Surfaces to Ore Evaluation." *Proc. Symposium on Mathematical Statistics and Computer Applications in Ore Evaluation,* Johannesburg, 13-38 (1966).

37. Morrison, D. F. *Multivariate Statistical Methods.* McGraw-Hill, New York (1967) 338 pp. (Excellent multivariate text.)

38. Anderson, T. W. *An Introduction to Multivariate Statistical Analysis.* John Wiley, New York (1958) 374 pp.

39. Kendall, M. G. *A Course in Multivariate Statistics.* Chas. Griffen London (1965) 185 pp. (Inexpensive paperback.)

40. Tryon, R. C. and D. E. Bailey, *Cluster Analysis.* McGraw-Hill, New York (1970) 347 pp.

41. Cattell, R. B. "Factor Analysis, an Introduction to Essentials." *Biometrics,* **21**, 190-215, 405-435 (1965).

42. Lawley, D. N. and A. E. Maxwell, *Factor Analysis as a Statistical Method.* Butterworth, London (1963).

BIBLIOGRAPHY

Each of the References cited previously is relevant as recommended reading to one or more of the subjects discussed in this chapter. Some additional items have been incorporated. The areas of relevance are suggested in the following approximate categories:

A. Selected references from various disciplines:

References 1, 9, 14, 15, 19 and the following:

B-1. Berry, B. L. J. and D. F. Marble *Spatial Analysis, A Reader in Statistical Geography*. Prentice-Hall Inc., Englewood Cliffs, New Jersey (1968) 512 pp. (A collection of definitive articles, essential reading for beginners and advanced workers alike.)

B-2. Cole, J. P. and C. A. M. King *Quantitative Geography*. John Wiley, New York (1968) 692 pp. (An overview of spatial analysis.)

B-3. Haggett, P. *Locational Analysis in Human Geography*. St. Martins Press, New York (1965) 339 pp.

B-4. Helava, U. V. *Some Trends in Automation of Photogrammetry*. Bildmessung u. Luftbildwesen **37**(6) (1969).

B-5. Kerr, A. J., T. V. Evangelatos, and D. Marshall "A System for Processing Hydrographic Data." *International Hydrographic Review* (1969).

B-6. Siegal, S. *Nonparametric Statistics for the Behavioral Sciences*. McGraw Hill, New York (1956) 312 pp. (An introductory text.)

B. Selected references on specific topics:

References 2-8, 10-13, 16-18, 20-42 inclusive and the following:

B-7. Cox, D. R. and H. D. Miller *The Theory of Stochastic Processes*. John Wiley, New York (1965) 398 pp. (Good on Markov chains.)

B-8. *Skylab Earth Resources Data Catalog*, JSC 09016/ Supt. of Documents, Washington, D.C. Stock No. 3300-00586 (1978).

B-9. Hebin, O. "Computer Drawn Isarithmic Maps." *Geografisk Tidsscrift* **68**(2), 50-63 (1969).

B-10. Koch, G. S., Jr., and R. F. Link, *Statistical Analysis of Geological Data*, Vol. 2. John Wiley, New York (1971) 438 pp. (Multivariate statistical analysis.)

B-11. Nordbeck, S. and P. Rystedt "Computer Cartography Point in Polygon Programs." *Lund Studies in Geography, Series C, General and Mathematical Geography*, No. **7**. (1967).

B-12. Rayner, J. N. *An Introduction to Spectral Analysis.* Pion, London. (1971).

B-13. Zzhcevm, Hr. A. and L. A. Mukina *Primenenie cuetoj i spektrozonal inoj aerofotos emki v geologicheskikh celjakh.* Izdatel. Moskoukogo Universitet, Moscow (1966). 303 pp.

Data Handling for Science and Technology
S.A. Rossmassler and D.G. Watson (eds.)
North-Holland Publishing Company

ANALYSIS AND INTERPRETATION OF DATA

G. D. James

ABSTRACT

Science and technology advance by the comparison of data with predictions and by the evaluation of data using all the information available. Some of the statistical techniques and concepts involved in this work are introduced in this chapter. These include mean, outliers, robust statistics, Student's t-test, the combination of errors, methods of parameter estimation -- the maximum likelihood techniques, Bayes theorem, and the method of least squares -- as well as non-parametric statistics.

The analysis and interpretation of data pervades science and technology. It includes the need to assign a correctly assessed error to a single datum derived by measurement, the determination of the best function and parameters to represent a relationship between variables, the need to detect and eliminate bias and spurious effects introduced by the instrumentation or the method of performing an experiment, and the testing of hypotheses and statistical inference. Many scientists accept the dictum that science proceeds by a series of conjectures and refutations (1) and that the distinguishing mark of a scientific statement is that it can be tested experimentally (2). The testing of hypotheses lies at the core of the method by which science advances. The hypotheses to be tested will at times involve a crucial decision in favour of one fundamental theory over another but more often will have more limited aims such as whether an instrument in use at one laboratory gives results in agreement with those from other instruments or whether the material produced by a given process has the properties that are demanded of it. Answers to questions like these are achieved in terms of properly assessed probabilities by the analysis and interpretation of data by statistical methods.

This chapter gives a brief outline of a small range of statistical methods as a means of referring the reader to more adequate and detailed publications. Statistical methods of experimental design (3-10) will not be described but it must be emphasised that an application of these techniques even to the simplest experiments can result in a far better understanding of the errors and limitations of the measurements performed. These methods also ensure that adequate data are obtained to answer the questions that the experiments set out to answer, and that adequate data are obtained to assign correct errors to the results obtained. The assignment of correctly measured errors to the quoted result of a measurement is of prime importance in using the result for data comparison, for data evaluation with the aim of deriving a "best value" from many measurements, and for comparison with theoretical predictions by the methods of statistical inference. Differences between repeated measurements of the same quantity under the same experimental conditions arise usually as the result of the sum of a number of small unperceived changes and, as a consequence of the central limit theorem, obey a Gaussian or normal probability distribution. The theory of the normal distribution and of the distribution of quantities which are functions of samples drawn from a normally distributed population (sampling theory) is thus of great importance in the analysis and understanding of errors of measurement. It should be remembered, however, that many quantities which arise in science are not themselves normally distributed. The basic methods of statistical analysis can take any known distribution into account. Also, robust techniques (11) have been developed which are insensitive to deviations from normality. In cases where the probability distribution of the data is unknown or difficult to handle, it is often possible to draw valuable conclusions by the use of distribution-free statistics (12).

In order to discuss methods of data analysis and interpretation by statistical means it is necessary to introduce a few statistical concepts and thereby introduce a number of statistical terms. If a measurement of a quantity x is made n times under "identical" conditions, the values obtained are said to constitute a sample. Conceptually, if the measurements were repeated indefinitely we would obtain the population of all possible measurements and be able to ascertain the population mean m which is alsc called the underlying mean or limiting mean. We would also generally observe that the population of measurements is distributed about the limiting mean with a Gaussian probability distribution characterised by a standard deviation σ such that in the units of x the distance from the mean value to the point of inflexion of the Gaussian distribution is equal to σ. The quantity σ^2 is known as the variance. Roughly 68.3 per cent of the population lies within $\pm\sigma$ of the limiting mean. The standard deviation σ is also equal to the root mean square of the amounts by which the measurements differ from m. These deviations from the limiting mean observed in repeated measurements under identical conditions are known as random errors and the standard deviation σ is a measure of the precision (or imprecision) of the measurements.

There is another type of error, known as the systematic error, which is equally important and not so easy to ascertain. Systematic errors remain constant over the period of the experiment and represent the difference between the population mean m and the "true" value of the quantity under investigation. This is also called the bias of the measurement. It often happens when measurements of a given quantity, made by different methods or at different times or in different laboratories, are compared, that they differ by much more than the errors ascribed. From an analysis of several important measurements, Youden (13) concludes that researchers often make little or no effort to ascertain the systematic errors in their measurements and that the errors quoted do not adequately reflect these errors. He also maintains that the only worthwhile estimates of systematic errors are those which are made experimentally. Systematic errors can be revealed by varying as many aspects of an experiment as possible. Often the things or quantities changed will have only two descriptions or values (for example carrying out a timing measurement over two separately measured paths instead of only one) but the benefit in bringing systematic errors to light will be immense. By careful planning, the time devoted to the experiments need not increase inordinately and it has been shown that a large reduction in the error in the mean can be achieved by changing more than one thing at a time. a large reduction in the error in the mean can be changing more than one thing at a time. Many papers describing how such experiments should be planned are now available, (See, e.g. (14).

The population mean m and the imprecision of the measuring process, σ cannot be determined exactly, but from the data available in a sample of size n we can derive estimates for these quantities. The sample average $\bar{x}$ may be used as an estimate of the population mean m and the sample standard deviation:

$$S = \left[\frac{\Sigma(x-\bar{x})^2}{n-1} \right]^{\frac{1}{2}}$$

may be used as an estimate of the imprecision σ. In this expression $(n - 1)$ represents the number of degrees-of-freedom available for estimating the error. In most cases the number of degrees-of-freedom associated with the sample standard deviation S is simply the number of measurements made less the number of parameters, derived from the data, which are used in the definition of S. The above expression for S contains one such parameter, $\bar{x}$, and thus the number of degrees-of-freedom associated with S in this case is $(n-1)$. The standard deviation of the mean, $\sigma_{\bar{x}}$, can be derived from the formula for the combination of errors to be presented later. For a set of n measurements drawn from the same population with standard deviation σ, it can be shown that the standard deviation of the mean is less than the standard deviation of an individual measurement by a factor $\sqrt{n}$. It is equal to $\sigma_{\bar{x}} = \sigma/\sqrt{n}$ and is estimated by $S/\sqrt{n}$. Using multiples of the standard deviation σ a confidence interval may be set up within which a chosen fraction of the data obtained will lie. For samples drawn from a normal population 95 per cent of the values of x lie within the interval $m \pm 1.96\,\sigma$. Similarly 95 per cent of the values of $\bar{x}$ lie within the interval $m \pm 1.96\,\sigma/\sqrt{n}$. To set up a confidence interval for m based on $\bar{x}$ requires a determination of the distribution of $|(x-m)/(S/\sqrt{n})|$ which is known as "Student's" t-distribution. It is tabulated (15) as a function of the number of degrees-of-freedom ν and the confidence interval $(1-\alpha)$. Using suffixes to indicate the specific dependence on n and ν, a confidence interval for m is given by

$$\bar{x}_n \pm t_\nu S_\nu/\sqrt{n}\,.$$

For example at the 95 per cent confidence level and $\nu = 5$ we have $t = 2.57$.

Another example of the use of "Student's" t-statistic is in the comparison of two means $\bar{x}$ and $\bar{y}$ derived from samples of size n and k from the same population. A pooled sample variance S_p^2 with $n+k-2$ degrees-of-freedom is derived by weighting the two sample variances S_x^2 and $S_{\bar{y}}^2$ by the degrees-of-freedom $(n-1)$ and $(k-1)$. Then, the quantity

$$t = \left[(\bar{x} - \bar{y}) - 0)\right]\left[(\frac{n+k}{nk})\right]^{-\frac{1}{2}} S_p$$

is distributed as "Student's" t-statistic. A confidence interval about $(\bar{x}-\bar{y})$ can be set up with $\nu = n+k-2$ degrees-of-freedom and confidence level $(1-\alpha)$. If this interval does not include zero we may conclude that the evidence is against the hypothesis that $m_{\bar{x}} = m_{\bar{y}}$.

The error quoted on the result of a given experiment is usually derived by combining the errors in the several quantities which enter into the expression for the final result. This can be done if the errors involved are small. It is easiest to proceed by combining the errors in the several quantities which enter into the expression for the final result. This can be done if the errors involved are small. It is easiest to proceed by combining the errors in two variables at a time. The general expression for the error $\sigma_{\overline{w}}$ in a function $\overline{w} = f(\overline{x}, \overline{y})$ of two variables is

$$\sigma_w^2 = \left[\frac{\delta f}{\delta x}\right]^2 \sigma_x^2 + \left[\frac{\delta f}{\delta y}\right]^2 \sigma_y^2 + 2 \left[\frac{\delta f}{\delta x}\right]\left[\frac{\delta f}{\delta y}\right] \rho_{\overline{xy}} \sigma_{\overline{x}} \sigma_{\overline{y}}$$

where the partial derivatives in the square brackets are to be evaluated at the average values of x and y. If $\overline{x}$ and $\overline{y}$ are independent, the correlation coefficient ρ_{xy} is zero. The variances $\sigma_{\overline{x}}^2$ and $\sigma_{\overline{y}}^2$ are estimated by $S_{\overline{x}}^2$ and $S_{\overline{y}}^2$ whereas the covariance $\rho_{\overline{xy}} \sigma_{\overline{x}} \sigma_{\overline{y}}$ is estimated by

$$S_{\overline{xy}} = n^{-1} (n-1)^{-1} \left[\Sigma (x - \overline{x})(y - \overline{y})\right] .$$

The error in some simple functions of $\overline{x}$ and $\overline{y}$ with $\rho_{xy} = 0$ have been tabulated by Ku (16). Let us assume that a quantity is measured and an error α_1 (the 'internal' error) is ascribed to the combination of errors as indicated above. Suppose now that the measurement is repeated n times with an independent assessment made of α_i each time. From the spread of the measurements it is also possible to ascribe an error α_i (the 'external' error) to each measurement. If the errors have been properly assessed, the ratio α_e / α_i will be unity with a standard error (17) given by $1/\sqrt{2(n-1)}$. This provides a useful check that errors are being properly assessed.

An important function of the analysis of data is to provide best estimates of quantities on which the observed data depend. Powerful statistical methods dominated by Bayes Theorem and the method of maximum likelihood have been developed to carry out this task (18). These two methods are not always equivalent and to some extent they reflect the difference between defining probability as a degree of belief and as a limiting frequency (19). The former definition results in the so-called Bayesian use of Bayes Theorem (20). Both methods can, however, be used with probability defined as a limiting frequency. A more familiar method of estimating the best parameters is the method of least squares first formulated by Legendre. This states that the most probable value of an observed quantity is such that the sum of the squares of the deviations of the observations from this value is least. When the data are normally distributed, the results obtained by least squares are identical with those derived by maximum likelihood. For data which do not have a normal distribution the method of least squares should be used with caution, although it often gives good results.

Next to the determination of the best value for a single parameter, one of the most useful applications of statistical analysis is in model fitting which seeks the best representation of the measured dependence of one variable on one or more other variables. The dependence of one variable on another can be either functional, such as the dependence of the density of a liquid on temperature, or statistical, such as the dependence of the age of a man on the age of his wife. An important first step in the analysis of such data is to make graphical plots to decide whether the relationship between two variables is reasonably linear or whether a more complicated model is necessary. It is often possible to linearise the relationship between two variables by, for instance, taking squares or logarithms. A detailed discussion of the methods involved is given by Natrella (21). A simpler account giving all the equations derived by least squares analysis is given by Topping (22). If the model adopted entails a power series dependence of one variable on another, the least squares estimation of the coefficients in the series results in a set of linear equations which can be solved explicitly (23). When a solution has been obtained it is instructive to plot the residuals to ensure that the model adopted is adequate. Statistical tests such as the likelihood ratio test can be used to determine how many terms in the series are justified by the data. For more complicated functions, efficient minimization computer programmes are available and can be adapted to estimate parameters by the least squares or maximum likelihood method (24).

It is sometimes found that one or more measured data points lie at a distance of several standard deviations from the predictions made by the model using the best estimated parameters. Such results are referred to as outliers or wild observations and systematic methods are required for their recognition and treatment (25,26). One approach is to replace each outlier by the result nearest to it in value but which is regarded as acceptable. Another is to set wild observations aside, as in the case of a trimmed mean, and study them separately. Such treatment is open to criticism and must be shown to be free from any attempt to achieve a desired result. A more readily acceptable approach is to use a statistic which is insensitive to deviations from normality in the distribution of the population. These are known as robust statistics (27-29).

Broadly speaking, tests on population means--such as "Student's" t-test for the mean of a normal population--are insensitive to departures from normality whereas tests on variance, such as the F-test for the ratio of two normal population variances, are very sensitive to such departures. Several tests, known as distribution-free or non-parametric tests, have been developed which are independent of the distribution of the population (30). For example, to test whether two sets of data have the same distribution and variance, a list of both sets of data (after normalisation to a common mean) is set out in numerical order. This list is transformed into a sequence of zeroes (designating the results from one data set) and ones (designating the results from the other data set). A run is then defined as a sequence of zeroes (or ones) terminated by ones (or zeroes). It has been shown that the distribution of the number of runs in the list is normal and expressions are available for the expectation value and variance (31). This runs statistic can be adapted (32) to search for structure in data where the underlying distribution is not Gaussian. In such work some insight is required to decide how best to translate the data from a numeral to an ordinal scale. It is also necessary to ascertain the effectiveness of the test selected in detecting the kind of structure expected. If the properties of the structure are known, so that simulated data sets, with various intensity of structure, can be generated using random numbers (33-34), the effectiveness of a selected test can be directly demonstrated.

Often in technology a large body of data is required for the successful operation of a complicated system or plant. To predict and control the behaviour of the system, data on the properties of material from which the system is constructed must be used in conjunction with data on the behaviour of the system. As time goes on, new data become available both on plant performance and also from measurements made elsewhere on the properties of materials. Periodically it becomes necessary to evaluate all the new data available and ensure that any adjustments made to data sets are consistent with the operational characteristics of the plant which themselves are, of course, subject to errors of measurement. Self-consistent evaluations can be made, using a covariance matrix to represent the interdependence of the measurements, by an application of Bayes' Theorem to connect the previous data set, which gives the 'prior' probabilities, with the new measurements, to derive a new data set characterised by the 'posterior' probabilities. It has been shown for instance that methods of nuclear data evaluation for nuclear reactor design developed independently in several countries are all equivalent to the above procedure (36).

REFERENCES

1. Popper, K. R. *Conjectures and Refutations*, Routledge and Kegan Paul, London (1969).

2. Popper, K. R. *The Logic of Scientific Discovery*, Hutchinson, London (1959).

3. Fisher, R. A. *The Design of Experiments*, 6th Ed., Oliver and Boyd, London (1960).

4. Cochran, W. G. and G. M. Cox *Experimental Designs*, John Wiley, London and New York (1957).

5. Davies, O. L. *The Design and Analysis of Industrial Experiments*, Oliver and Boyd, London (1954).

6. Kempthorne, O. *The Design and Analysis of Experiments*, John Wiley, New York (1952).

7. Fisher, R. A. and F. Yates *Statistical Tables for Biological, Agricultural and Medical Research*, Oliver and Boyd, Edinburgh and London, Hafner, New York (1963).

8. Snedecor, G. W. and W. G. Cochran *Statistical Methods* 6th Edition, Iowa State University Press, Ames, Iowa (1967).

9. Winer, B. J. *Statistical Principles in Experimental Design*, 2nd Edition, McGraw-Hill, New York (1971).

10. Quenehen, A. "Techniques statistiques de traitement des données," Communication au colloque CODATA-France *Utilisation de données-Banques de données*, Paris, (Sept. 1977).

11. Box, G. E. P. "Non-normality and tests on variance," *Biometrika* **40**, 318 (1953).

12. Fraser, D. A. S. *Non-parametric Methods in Statistics*, Wiley, New York (1957).

13. Youden, W. J. "Enduring values," *Technometrics* **14**, 1 (1972).

14. Ku, H. H., ed. *Precision Measurement and Calibration*, NBS Special Publication 300-Vol. 1., Statistical Concepts and Procedures, Washington, D.C. (1969).

15. Pearson, E. S. and H. O. Hartley, *Biometrika Tables for Statisticians*, 3rd Edition, Vol. I, The University Press, Cambridge (1966) p. 146.

16. Ku, H. H., "Notes on the use of propagation of errors formulas," loc. cit., p. 331.

17. Topping, J. *Errors of Observation and their Treatment*, 3rd Edition, Chapman and Hall, London (1962) p. 43.

18. Kendall, M. G. and A. Stuart, *The Advanced Theory of Statistics*, Vol. I, Charles Griffin, London (1958) p. 198.

19. von Mises, R. *Probability, Statistics and Truth*, Wm. Hodge, London (1938).

20. Eadie, W. T., D. Gryard, F. E. James, M. Ross and B. Sadoulet, *Statistical Methods in Experimental Physics*, North Holland, Amsterdam (1971) p. 13.

21. Natrella, M. G. "Characterizing linear relationships between two variables." reprinted in Ku, loc. cit., p. 204.

22. Topping, J., loc. cit., p. 96.

23. Wampler, R. H. "Some recent developments in linear least squares computations," *Proceedings, Computer Science and Statistics, Sixth Annual Symposium on the Interface*, M. E. Tarter, ed., University of California, Berkeley (1972).

24. Fletcher, R. "Function minimization without evaluating derivatives, a review," *Computer J.* **8**, 33 (1965).

25. Kruskal, W. H. "Some remarks on wild observation," reproduced in Ku, loc. cit., p. 346.

26. Proschan, F. "Rejection of outlying observations," reproduced in Ku, loc. cit., p. 349.

27. Box, G. E. P. and S. L. Anderson "Permutation theory in the derivation of robust criteria and the study of the departures from assumption," *J. Roy. Statistical Soc.* **B17**, 1 (1955).

28. Tukey, J. W. "The future of data analysis," *Ann. Math. Statistics* **33**, 1 (1962).

29. Huber, P. J. "A robust version of the probability ratio test," *Ann. Math. Statistics* **36**, 1973 (1965).

30. David, H. A. *Order Statistics*, John Wiley, New York (1970).

31. Wald, A. and J. Wolfowitz "On a test whether two samples are from the same population," *Ann. Math. Statistics* **11**, 147 (1940).

32. James, G. D. "Application of distribution free statistics to the structural analysis of slow neutron cross section and resonance parameter data," *Nucl. Phys.* **A170**, 30 (1971).

33. Tocher, K. D. *The Art of Simulation*, The English University Press Ltd., London (1963).

34. Hammersley, J. M. and D. C. Handscomb *Monte Carlo Methods*, Methuen, London (1964).

35. Meyer, H. A., ed. *Symposium on Monte Carlo Methods*, John Wiley, New York (1956).

36. Dragt, J. B., J. W. M. Dekker, G. Gruppelaar and A. J. Janssen "Methods of adjustment and evaluation of neutron capture cross sections," *Nucl. Sci. and Eng.* **62**, 117 (1977).

BIBLIOGRAPHY

B-1. Acton, F. S. *Analysis of Straight Line Data*, John Wiley, New York (1959).

B-2. Bendat, J. S. and A. G. Piersol *Measurement and Analysis of Random Data*, John Wiley, New York (1966).

B-3. Box, G. E. P. and G. C. Tiao *Bayesian Inference in Statistical Analysis*, Addison-Wesley, Reading, Massachusetts (1973).

B-4. Box, G. E. P. and G. M. Jenkins *Time Series Analysis, Forecasting and Control*, Holden-Day, San Francisco, California (1970) Revised 1976.

B-5. Brownlee, K. A. *Statistical Theory and Methodology in Science and Engineering*, 2nd Edition, John Wiley, New York (1965).

B-6. Cooper, B. E. *Statistics for Experimentalists*, Pergamon Press, Oxford (1969).

B-7. David, H. A. *Order Statistics*, John Wiley, New York (1970).

B-8. Eadie, W. T., D. Brijard, F. E. James, M. Roos and B. Sadoulet *Statistical Methods in Experimental Physics*, North Holland, Amsterdam (1971).

B-9. Fisher, R. A. *Statistical Methods for Research Workers*, 14th Edition, Oliver and Boyd, Edinburgh (1970).

B-10. Hoel, P. G. *Elementary Statistics*, 4th Edition, John Wiley, New York (1976).

B-11. Hoel, P. G. *Introduction to Mathematical Statistics*, 3rd Edition, John Wiley, New York (1962).

B-12. Hollander, M. and D. A. Wolf *Non-parametric Statistical Methods*, John Wiley, New York (1973).

B-13. Kempthorne, O. and J. L. Folks *Probability, Statistics and Data Analysis*, Iowa State University Press, Ames, Iowa (1971).

B-14. Kendall, M. G. and A. Stuart *The Advanced Theory of Statistics*,

Vol. 1. *Distribution Theory*, 3rd Edition (1969).
Vol. 2. *Inference and Relationship*, 3rd Edition (1973).
Vol. 3. *Design and Analysis, and Time Series*, 2nd Edition (1968). Griffin, London.

B-15. Ku, H. H. *Statistical Concepts and Procedures*, NBS Special Publication 300, Vol. 1, Washington, D.C. (1969).

B-16. Lehman, E. L. *Testing Statistical Hypothesis*, John Wiley, New York (1959).

B-17. Mandel, J. *The Statistical Analysis of Experimental Data*, John Wiley, New York (1964).

B-18. Moroney, M. J. *Facts from Figures*, Penguin Books Ltd., Harmondsworth (1954).

B-19. Parrott, L. G. *Probability and Experimental Errors in Science*, John Wiley, New York (1961).

B-20. Pugh, E. M. and G. W. Winslow *The Analysis of Physical Measurements*, Addison-Wesley, Reading, Massachusetts (1966).

B-21. Tukey, J. W. *Exploratory Data Analysis*, Addison-Wesley, Reading, Massachusetts (1976).

B-22. Youden, W. J. *Statistical Methods for Chemists*, John Wiley, New York (1951).

B-23. Yule, G. U. and M. G. Kendall *An Introduction to the Theory of Statistics*, 14th Edition, Charles Griffin, London (1958).

B-24. Weatherburn, C. *A First Course in Mathematical Statistics*, The University Press, Cambridge (1952).

Data Handling for Science and Technology
S.A. Rossmassler and D.G. Watson (eds.)
North-Holland Publishing Company

PRESENTATION OF DATA IN THE PRIMARY LITERATURE

S. A. Rossmassler

ABSTRACT

This chapter examines the primary scientific literature as a channel for the communication of data. Problems in presenting data in the primary literature, as encountered both by authors and by readers, are identified. Solutions and ameliorations of these problems are noted.

INTRODUCTION

The "primary literature" as considered here includes all of the various general and specialized technical and professional journals in which scientific workers, engineers, field observers, etc., publish the results of their own research and observation. The inclusion of technical reports and patents as part of the primary literature raises questions which are discussed below.

The historical role of the primary literature as the major (primary!) channel for communicating results to interested readers is well established. The importance of such communication is clearly stated in a report by a CODATA Task Group (1). "The remarkable progress of natural science has been made possible by virtue of the systematic structure of scientific knowledge: namely, new research results are built upon the achievements of predecessors, and this successive accumulation of knowledge has led to the construction of the huge edifice of modern science. The essential factor in this construction process of science is the smooth communication of research results, as scientific information, among research workers. The applicability of the achievements of science to technology also depends on this feature of scientific knowledge." Further examination of the motivations of originators of technical papers, as well as various aspects of publishing scientific journals, is provided in the SATCOM Report (2); also in a review by Gannett (3). Additional sources are identified in Category A of the chapter bibliography.

PROBLEMS IN DATA PRESENTATION

The first difficulty with the primary literature as a source of information, and more specifically as a source of data, is the volume of literature published. Perhaps the clearest analysis of this paradox is presented by Simon (4) in his chapter "Designing Organizations for an Information-Rich World." As he states, "a wealth of information creates a poverty of attention and a need to allocate that attention among the overabundance of information sources that might consume it." Weinberg (5) describes the situation in different words, "The capacity to absorb information limits the system." There is ample statistical evidence of the continuing growth of the scientific literature. The U.S. National Federation of Abstracting and Indexing Services estimates close to one million articles abstracted in 1974 by its member societies. The growth rate of scientific literature appears to have been rising on a curve which approximates 7% per year, although recent statistics suggest that the rate may at present be only one-half that value. Projections for future growth and interpretations of the impact and implications are given by Anderla (6) in a study made for the Organization for Economic Cooperation and Development.

This situation leads to several specific problems. First, the volume of material presented for publication strains the capacity of the journals. This strain leads editors to impose restrictions on the number of pages which the author can devote to comprehensive presentation of his data results -- frequently only summary tables, examples of graphs, etc., can be included. Equally serious is the fact that full exposition of measurement conditions, calibrations, etc., must often be sacrificed. As a result, only a fraction of the data are immediately accessible to the user, and those data which are presented cannot, in many cases, be adequately interpreted, analysed, or compared to other results. The reader is free to write to the author, of course, and this channel of communication is sometimes quite rewarding. However, even at best it involves delays (particularly when the mail must go from one country to another) and substantial costs if many pages of typewritten material are involved. If the paper is more than a few years old, the author may have stored or even lost his original material.

A second difficulty is that the primary literature is, to a large extent, a peer-group communication device (see Herschman, 7). A worker in a frontier field does not always feel obliged to explain to non-expert readers the meaning or breadth of application of his results. Consequently, possible useful data are not even identified as such by the author or considered for use by the data seeker in many problem-solving situations. With the increasing availability of computer-readable full text of journal articles (deriving from the growing use of computer typesetting), it is sometimes possible to overcome these difficulties. O'Connor (8) reports 60% success in retrieving data-containing papers in a recent experiment.

A more systematic solution is possible through the inclusion of data indicators in the abstract of each paper. Progress in this direction is described in Chapter 6, "Access to Data in the Primary Literature," of this ***Sourcebook***.

Other sets of difficulties affect the utility of the patent literature and the technical report literature. The patent literature is a valuable source of data, and helpful guides exist to finding patents relevant to any technical subject (9-17). However, the role of the patent application as a legal document, which may seek only to establish priority of invention, gives the applicant little or no incentive to present all of the data he has gathered. Only those data which define the invention and provide examples of its operability and scope are relevant to the process of obtaining a patent, and only such data will normally be included.

The technical report literature may be less than highly regarded as a peer-group communication channel among research workers, but its value in many applied research and development efforts cannot be denied. Technical reports frequently contain data of great value. Since space limitations do not usually apply, the author can include all of his data, and can describe fully the details of his observational procedures. The report literature is, however, less accessible (or less easily accessible) to the outside data user than is the journal literature. Several excellent analyses of the technical report literature have been presented (9,18-20).

Technical reports fall into two general categories — reports of work done under contract to a governmental body, and reports of work done in private laboratories. Reports in the latter category are seldom available to those outside the sponsoring organization (private corporation, privately funded research institute, etc.), and they will not be discussed further in this chapter. Internal and informal reports based on work done at universities, research institutes, etc., are frequently valuable sources of data. Such reports may receive wide dissemination, but are seldom available through formal channels. Governmentally sponsored technical reports are frequently available to the general public; in fact, a whole subuniverse of literature handling has developed, devoted to technical reports.

In the U.S.A., the National Technical Information Service (formerly the Clearinghouse for Federal Scientific and Technical Information) is the major central source for the public sale of Government-sponsored research, development, and engineering reports. The NTIS information collection includes nearly one million titles. Services include an on-line computer-based bibliographic search system, publication of a weekly abstract journal, selective dissemination of reports in microfiche format, etc. Details can be obtained from the NTIS, 5285 Port Royal Road, Springfield, Virginia 22151, U.S.A. It should be noted that NTIS provides documents containing data; it does not extract or compile data from such reports.

A somewhat parallel organization in the USSR is the National Information System for Research and Development (NIS R&D), a part of the USSR State Scientific and Technical Information System. The NIS R&D executive body is the All-Union Scientific and Technical Information Centre (VNTITsentr) under the State Committee of the USSR Council of Ministers for Science and Technology.

VNTITsentr is in official charge of keeping the state registry of initiated, current, and completed Research and Development and Demonstration and Design projects. Originally (from 1968) the Centre's terms of reference only embraced information on research in natural science, exact and technical sciences, medicine, and agriculture. Later (from 1969 on) it encompassed social sciences, and, since 1973, also design and development projects. So VNTITsentr now concerns itself with all fields of science and engineering, accumulating and disseminating information about both R&D and D&D projects. Since 1969, the Centre's scope widened to cover the upheld doctoral and candidate's dissertations.

The Centre provides services only to a limited group of corporate users -- councils on scientific and technical problems, a group of leading R&D institutes, and agencies responsible for management of scientific and technological progress.

The System Principal is the Director of the All-Union Scientific and Technical Information Centre (VNTITs). Their address is 125493, Moscow, Smolnaya 14, USSR.

Other national information programs for science and technology (see Appendix I) which focus to varying degrees on numerical data also provide systematic accessibility to the primary literature. For example, the Indian government, which sponsors a variety of data programs in science and technology, provides direct contact with world bibliographic data bases. The current status of this effort was described at the Sixth International CODATA Conference, held in Santa Flavia, Italy (21).

In addition to such centralized national services, some individual governmental agencies also maintain abstracting and indexing services for the report literature which they produce. Some of these services are available to the general public, while some are restricted to users having a formal affiliation with the sponsor. Overall, the technical report literature is a valuable information resource, but one which cannot be completely utilized without experience.

SOLUTIONS TO DATA PRESENTATION PROBLEMS

The preceding pages have identified some of the channels available for the presentation of data (and other research outputs) in original publications, and difficulties associated with those channels. Solutions to these various difficulties have been proposed, and many of the most appealing are being implemented. One general approach involves the information analysis center or data center, which is discussed in detail in the Weinberg report (5), in a recent Unesco report (22), in a book by Weisman (23), and in many other publications. Chapter 7, "Compilation and Evaluation of Data," in this ***Sourcebook*** discusses data centers at some length.

A partial solution to the problem of incompleteness of data as presented in the primary literature is the data depository, a supplementary access point which the author of an article in the journal literature may use. If some of his valuable data cannot be included in his primary publication, he can place a more complete file in such a depository, and make reference to their availability in the journal article.

Data depositories are maintained in the United States of America by the American Chemical Society Microfilm Depository Service, 1155 Sixteenth Street, Washington, DC 20036, by the American Institute of Physics (Physics Auxiliary Publication Service), 335 East 45th Street, New York, NY 10017, and by the National Auxiliary Publications Service of the American Society for Information Science (ASIS/NAPS), c/o Microfiche Publication, P. O. Box 3513, Grand Central Station, New York, NY 10017. The Canadian government maintains a Depository of Unpublished Data, CISTI, National Research Council of Canada, Ottawa, Canada K1A 052.

Another partial solution is the specialized data journal, a journal devoted primarily to data in a particular field. Such journals must be distinguished from review journals, and from reference data journals. The data journal accepts articles by individual authors, consisting wholly or largely of their own measurement results. Examples are *Atomic Data and Nuclear Data Tables*, published by Academic Press, and the *Journal of Chemical and Engineering Data*, published by the American Chemical Society. Review journals, such as *Reviews of Modern Physics*, and the Annual Reviews series, contain papers which are critical examinations of a particular field or topic, but which seldom stress data or numerical aspects except to illustrate a particular point. Reference data journals, on the other hand, contain compilations of data from a comprehensive range of primary journal articles. The data do not derive from a single laboratory, but from all relevant measurements. The author of the compilation exercises critical evaluation and selection procedures to determine the "best" (i.e., the most reliable) data. *The Journal of Physical and Chemical Reference Data*, published by the American Chemical Society and the American Institute of Physics for the U.S. National Standard Reference Data program, is the most familiar current example of a reference data journal.

Responsibility for improving the presentation and availability of data cannot be assigned totally to any single group. For example, the effectiveness of data presentation in the primary literature can be substantially increased through the author's adherence to several sets of instructions which have been prepared for that purpose. Most generally applicable is the Guide prepared by the CODATA Task Group on Publication of Data in the Primary Literature (24). Other useful guidelines (25-31) are available on more specific categories of data. Chapter 8 of the ***Sourcebook*** also discusses guidelines, standards, and other criteria for the publication of data. In addition, a fair share of responsibility for finding data must remain with the data user. Data users (and, in fact, users of all kinds of information) are typically reluctant to seek out new or improved channels for acquiring data which they need.

SPECIALIZED DATA PUBLICATIONS

A comprehensive list of numerical data sources is maintained by CODATA, and updated publications of the list are provided from time to time. The World Data Referral Centre, with sponsorship by both CODATA and Unesco, affords identification of specialized and broad-coverage data publications (see Chapter 10 for further discussion of these resources).

REFERENCES

1. "Study on the Problems of Accessibility and Dissemination of Data for Science and Technology." CODATA Task Group on Accessibility and Dissemination of Data. SC.74/WS/16. ***CODATA Bulletin*** No. **16** (October 1975).

2. *Scientific and Technical Communication. A Pressing National Problem and Recommendations for Its Solution.* National Academy of Sciences, Washington, D.C. (1969) (especially Chapter 4).

3. Gannett, F. K. "Primary Publication Systems and Services." Chapter 8 of *Ann. Rev. Inform. Sci. Tech.*, Vol. 8. C. A. Cuadra, ed., A. W. Luke, assoc. ed., American Society for Information Science, Washington, D.C. (1973).

4. Simon, H. A. "Designing Organizations for an Information-Rich World." Chapter 2 of *Computers, Communications and the Public Interest.* M. Greenberger, ed., Johns Hopkins University Press, Baltimore, Maryland (1971).

5. *Science, Government and Information. The Responsibilities of the Technical Community and the Government in the Transfer of Information.* A Report of the President's Science Advisory Committee. US Government Printing Office, Washington, D.C. (1963).

6. Anderla, G. *Information in 1985. A Forecasting Study of Information Needs and Resources.* Organization for Economic Cooperation and Development, Paris (1973).

7. Herschman, A. "The Primary Journal: Past, Present, and Future." *J. Chem. Documentation* **10**, 1 (February 1970).

8. O'Connor, J. "Data Retrieval by Text Searching." *J. Chem. Inform. Computer Sci.* **17**, 181 (August 1977).

9. Gould, R. F., ed. *Searching the Chemical Literature.* Advances in Chemistry Series 30, American Chemical Society, Washington, D.C. (1961).

10. Cole, R. I., ed. *Data/Information Availability.* Thompson Book Company, Washington, D.C. (1966).

11. McDonnell, P. M. "Searching for Chemical Information in the Patent and Trademark Office." *J. Chem. Inform. Computer Sci.* **17**, 122 (August 1977).

12. Maynard, J. T. "Chemical Abstracts as a Patent Reference Tool." *J. Chem. Inform. Computer Sci.* **17**, 136 (August 1977).

13. Donovan, K. M. and B. B. Wilhide. "A User's Experience with Searching the IFI Comprehensive Database to U.S. Chemical Patents." *J. Chem. Inform. Computer Sci.* **17**, 139 (August 1977).

14. Kaback, S. M. "A User's Experience with Derwent Patent Files." *J. Chem. Inform. Computer Sci.* **17**, 143 (August 1977).

15. Smith, R. G., L. P. Anderson, and S. K. Jackson. "On-Line Retrieval of Chemical

Patent Information. An Overview and a Brief Comparison of Three Major Files." *J. Chem. Inform. Computer Sci.* **17**, 148 (August 1977).

16. Pilch, W. and W. Wratschko. "INPADOC: A Computerized Patent Documentation System. *J. Chem. Inform. Computer Sci.* **18**, 69-75 (1978).

17. Marcus, M. J. "Patents and Information." *J. Chem. Inform. Computer Sci.* **18**, 76-78 (1978).

18. Passman, S. *Scientific and Technological Information.* Pergamon Press, London (1969).

19. Herner, S. *A Brief Guide to Sources of Scientific and Technical Information.* Information Resources Press, Washington, D.C. (1969).

20. *The Role of the Technical Report in Scientific and Technological Communication.* Prepared by COSATI Task Group on the Role of the Technical Report, A. A. Aines, Chairman. (December 1968). (Available from NTIS, Washington, D.C. as PB 180 *944.)*

21. *Proceedings of the Sixth International CODATA Conference,* Santa Flavia, Palermo, Italy, 22-25 May, 1978. Pergamon Press, Oxford and New York (March 1979).

22. *UNISIST Guidelines for Establishing and Operating Information Analysis Centers.* Prepared with the collaboration of F. Kertesz, Unesco, Paris (1978).

23. Weisman, H. M. *Information Systems, Services and Centers.* Becker and Hayes, Inc., New York (1972).

24. "Guide for the Presentation in the Primary Literature of Numerical Data Derived from Experiments." Report of the CODATA Task Group on Publication of Data in the Primary Literature. ***CODATA Bulletin*** No. **9** (December 1973).

25. "The Presentation of Chemical Kinetics Data in the Primary Literature. Report of the CODATA Task Group on Data for Chemical Kinetics." ***CODATA Bulletin*** No. 13 (December 1974).

26. "Specifications for Evaluation of Infrared Reference Spectra." The Coblentz Society Board of Managers. *Anal. Chem.* **38**, 27A-38A (June 1966).

27. "The Coblentz Society Specifications for Evaluation of Research Quality Analytical Infrared Spectra (Class II)." *Anal. Chem.* **47**, 945A-952A (September 1975).

28. "Recommendations for Data Compilations and for Reporting of Measurements of the Thermal Conductivity of Gases." *J. Heat Transfer*, 479-480 (1971).

29. "Recommendations for the Presentation of NMR Data for Publication in Chemical Journals. IUPAC Physical Chemistry Division Commisssion on Molecular Structure and Spectroscopy." *Pure Appl. Chem.* **29**, 627-628 (1972).

30. Bartell, L. S., K. Kuchitsu, and H. M. Seip. "Guide for the Publication of Experimental Gas-Phase Electron Diffraction Data and Derived Structural Results in the Primary Literature." *Acta Cryst.* **A32**, 1013-1018 (1976).

31. "Recommendations for the Presentation of Raman Spectral Data." NRC-NAS Ad Hoc Panel: Raman Spectral Data. *Appl. Spect.* **30**, 20-23 (1976).

BIBLIOGRAPHY

B-1. Kochen, M., ed. *The Growth of Knowledge. Readings on Organization and Retrieval of Information.* John Wiley, New York (1967).

B-2. *Report of the Task Group on the Economics of Primary Publication.* National Academy of Sciences, Washington, D.C. (1970).

B-3 *Information for a Changing Society, Some Policy Considerations.* Ad Hoc Group on Scientific and Technical Information, P. Piganiol, Chairman. Organization for Economic Cooperation and Development, Paris (1971).

B-4. Nelson, C. E. and D. K. Pollock, ed. *Communication Among Scientists and Engineers.* D.C. Heath, Lexington, Massachusetts (1970).

B-5. Price, D. J. *Little Science, Big Science.* Columbia University Press, New York (1963).

RECOMMENDED READING

Each of the References cited previously is relevant as recommended reading to one or more of the subjects discussed in this chapter. The areas of relevance are suggested in the following approximate categories:

A. General references on the scientific literature, scientific communication processes, etc.:

1-7, 9, 11, 13, 18, 19, 20, 22, 23, B-1, B-2, B-4, B-5

B. References on availability of data in the primary literature:

1, 5, 7, 10, 22, 23, B-2, B-3

C. References on technical reports and other special literature formats:

2, 3, 5, 9, 10, 18, 19, 20, B-2

D. References on data centers, information analysis processes, etc.:

1, 2, 5, 10, 22, 23, B-2, B-3

E. Specific guidelines, standards, and criteria for the presentation of data:

24-31

Data Handling for Science and Technology
S.A. Rossmassler and D.G. Watson (eds.)
North-Holland Publishing Company

ACCESS TO DATA IN THE PRIMARY LITERATURE

R. G. Lerner

ABSTRACT

Current experiments in developing data-descriptive records in the primary and secondary literature are described. Access to the primary literature through secondary services and acquisition of the full text of primary documents are discussed.

INTRODUCTION

Access to the primary literature has two facets: location of the bibliographic citation of primary literature containing data, and access to the actual documents. Methods for improving the presentation of data in the primary literature are discussed in Chapters 5 and 8. Since abstracts are generally required or prepared for scientific journal articles and reports, efforts to improve access to numerical data presented in the primary literature have focussed on data published in, or in conjunction with, the abstract. Standards for writing abstracts have been published by the American National Standards Institute (1) and elsewhere in an article by Weil (2) and by a Unesco guide for the preparation of authors' abstracts for publication (3). According to the Weil article, the abstract "should present as much as possible of the quantitative and/or qualitative information contained in the document." The Unesco guide points out that:

"New information should include observed facts, conclusions of an experiment or argument, and essential points of a new method of treatment or of newly designed apparatus, etc. When feasible, it is preferable to give specific numerical results rather than merely to say what was measured. Reference should be made to new material (compounds, etc.) and new numerical data, such as physical constants. Attention should be drawn to these even though they may be incidental to the main purpose of the paper. Otherwise valuable information may be hidden. When an abstract includes experimental results there should also be some indication of the method used. Reference to new methods should include their basic principle, the range of operation and the degree of accuracy of results."

DATA INDICATORS

Several studies have been conducted which are intended to develop systems for attaching data descriptive records to abstracts and bibliographic citations at the time of publication, in a manner which would be appropriate for, and compatible with, the various subdisciplines of science and engineering. The most promising approaches are based on data flagging and tagging in the primary and secondary literature (4). A data flag has been defined as a code symbol which indicates the presence of data and may -- but need not -- identify the type of data reported in a primary publication; for example, there could be one or more appropriate symbols at the end of an abstract in the primary publication which would have been established internationally and which would indicate the various types of data in that paper. A data tag is a more detailed identifier which is used to characterize scientific observations, measurements, or calculations on the substance or system; data tags can be regarded as "keywords" for data. It is hoped that this would allow identification of every abstract of an article or report that contains data, and a designation of the physical or biological quantities involved in the data. The relevance of these techniques to both personal and computerized searching for data is obvious.

The American Institute of Physics has studied ways in which primary publishers could assist data centers and individuals by indicating the data content of journal articles (5) at the time of publication. In collaboration with a group of data centers of the National Standard Reference Data System and the editors of some of the journals published by the Institute, a range of possibilities was explored:

- A simple notation that data occurs in the article; unfortunately, this would not show the type of data, and would still require reading the whole article to determine the relevance of the data contained.
- Detailed indexing for each data center, which would have been prohibitively expensive in terms of specialist personnel requirements.
- Collecting the captions of figures and tables in the article, which was rejected because of the ambiguity of the wording of captions.
- Use of data flags such as the IUPAC list (6), which lacked identification of materials and environmental conditions, and which uses a short mnemonic tag which could not be expanded to cover more than a small area of science and engineering.
- Expansion of the "notation of content" or "key phrases" concept which combines data flags and data tags; this system is presently being used in *Physical Review C* and in *Nuclear Physics* for articles in the field of nuclear physics.

The last alternative was felt to be the most flexible, and a structured format was worked out which could be used by an author to provide the data notation for his own paper. The format requires a data flag, chosen from a broad list of fixed-language terms, followed by free-language flags in ordered strings of material or system description, properties or parameters determined, relevant environment (mass, temperature or pressure ranges), and space for comments. The data centers felt that, with this published along with the printed abstract and the computer tape of the abstract, so that it was available for both human and computer scanning, any description of methods would be superfluous. A form was designed for use by authors, together with instructions for filling out the form; the project has not yet been tested in actual operation.

Chemical Abstracts Service (CAS) is conducting two data tagging experiments; one study is evaluating data tagging from a document-oriented point of view and the other from the more specific substance-oriented viewpoint within each document. The first tagging experiment (7) focuses on the biochemical subject area; tags will appear in a special computer-readable version of *Chemical-Biological Activities (CBAC)*. CAS is working with the National Library of Medicine, which will aid in the evaluation of the usefulness of the tags. The tags appearing in the special *CBAC* package will be of two types: data tags which are codes to indicate the presence of specific types of numerical data in a corresponding source document, and information tags which are codes to indicate certain conceptual subjects which would not normally appear in an abstract or index entry for a source document (an example would be "SAF" to show that safety information is contained somewhere in the document; this safety information may be mentioned only incidentally and therefore would not be brought to the user's attention by a secondary service). Both kinds of tags would be document-oriented; i.e., if infrared spectral data are found in a document, this will be tagged, and that tag will be associated with the document identification number. The second tagging experiment (8) uses the CAS computer-readable *Energy* service as a vehicle. A special issue of *Energy* will be prepared which contains tags; it will be evaluated by users at Oak Ridge National Laboratory. The information tagging scheme will be similar to that for *CBAC*; however, the data tags will be associated with specific chemical substances identified in the source document. Providing this level of detail will be costlier and more time-consuming than the *CBAC* data tagging, since each item of data will be separately identified and recorded. A second task in this tagging work will be a study of the types of data found in those sections of *Chemical Abstracts* which are not covered in this work, and an assessment of extending data tags to these other areas.

The National Aeronautic and Space Administration (NASA), with the assistance of the Denver Research Institute, is carrying out an operational test of numerical data tagging and flagging in NASA's scientific and technical information system (9). A sample of journal articles and technical reports in the field of metallic and nonmetallic materials and in an area of physics is being selected. The numerical data are flagged and tagged, and made accessible through NASA's secondary services, in printed and electronic forms. The test involves the real-time users of NASA's system through announcement in the standard abstract journals, *STAR* and *IAA*, and retrieval through the on-line interactive computer system, *NASA/RECON*. Feedback on newly announced and retrieved bibliographic citations containing appropriate data summaries will be compared with similar items not containing data summaries. The study is designed to determine the impact which the data summaries would have on the speed with which users receive pertinent data, the precision with which a user's requirements are satisfied, and the volume of usage of the numerical data content of the aerospace literature.

Under the direction of Engineering Index, an abstracting and indexing service, a project on accessing engineering numerical data is being carried out at the Purdue University Center for Information and Numerical Data Analysis and Synthesis (10). The study will gather and analyze the ways in which engineers use information/data systems: the study will also assess alternative services that may be derived from the CINDAS data base. CINDAS will prepare software to enable on-line interactive access to its data bank, convert portions of its evaluated data to machine-readable form to use in a test marketing effort, and provide technical support to the overall study. The system should permit an engineer to specify the parameters he wants his material to have, and to respond by making available a listing of the materials which satisfy those criteria.

The International Nuclear Information System (INIS) is also studying methods for including extra data tagging or flagging information with abstracts. Under their proposed system, an entry would carry a data flag "N" when data are present; new tagged fields with extra data indexing would contain information on the way in which the data were obtained and a limited number of descriptors (11).

The National Science Foundation has funded a study (12) by Informatics, Inc. to analyze the state of the art on indexing the numerical data contained in the scientific and technical literature; to identify the problems associated with data indexing and data accessibility; to assess the progress of data indexing in journals, abstracting services and information centers; and to develop conclusions that will serve as guidelines which will make numerical data more accessible to users. Another study reported by O'Connor (13) has focused on the effectiveness of data retrieval through whole-text searching.

ACCESS TO CITATIONS AND ABSTRACTS

Finding citations to appropriate literature can be accomplished in several ways: by using printed or computerized abstracting and indexing services, by relying on a computerized information center, or through the services of an information broker.

Abstracting and indexing services provide bibliographic aids in the form of indexes and brief descriptions of primary literature. Printed aids may take the form of abstract journals (some of which have specialized indexes), index publications which provide bibliographic data without abstracts, citation indexes which correlate journal articles with later references to the same research paper, and collections of tables of contents. Other products provided by abstracting and indexing services include magnetic tapes, microforms and punched cards (14,15).

Most of the abstracting and indexing services are discipline oriented. Each discipline may have several services, offering products in different languages. Many of these services belong to the International Council of Scientific Unions/Abstracting Board (ICSU/AB), which promotes cooperation and adoption of common practices and standards among the various national and disciplinary groups (16). A few abstracting and indexing services are interdisciplinary in nature, such as the International Nuclear Information System (INIS) of the International Atomic Energy Agency, which publishes *Atomindex* and *INIS* tapes (17), and the National Technical Information System in the United States, which publishes tapes and weekly technical report summaries (18). In the United States, most of the American abstracting and indexing services belong to the National Federation of Abstracting and Indexing Services (NFAIS). NFAIS has conducted a study of the overlap among its members, and also issues an annual report on the number of items covered by each service (19).

Many services offer their data bases on magnetic tape. Among the directories which summarize the available data bases are the *International Directory of Computer and Information System Services* (20), *Encyclopedia of Information Systems and Services* (21), *Computer-Readable Bibliographic Data Bases--A Directory and Data Sourcebook* (22), and the *ASIDIC Survey of Information Centers Using Machine-Readable Data Bases* (23). Further information on data bases, information centers and information brokers in the United States and Europe can be obtained from the meeting notes and newsletters published by the Association of Information Dissemination Centers (ASIDIC) (24) and European Scientific Information Dissemination Centers (EUSIDIC) (25). Annual surveys on the generation and use of machine-readable data bases have been prepared by Williams, Schipma and Wilde for the *Annual Reviews of Information Science and Technology* (26-28). A discussion of retrieval systems and techniques also appears in a review article by Summit & Firschein (29).

The *World Inventory of Abstracting and Indexing Services* (being prepared by FID and NFAIS with partial support from Unesco and NSF) will be available in machine-readable and printed form in 1979. Printed copies will be obtainable, at a price to be determined, through National Technical Information Service (NTIS), 5285 Port Royal Road, Springfield, Virginia 22151, USA. Queries concerning services to be available from the master tape file should be addressed to the International Federation for Documentation (FID), P.O. Box 30115, 2500 GC, The Hague, Netherlands, or National Federation of Abstracting and Indexing Services (NFAIS), 3401 Market Street, Philadelphia, Pennsylvania 19104, USA. The inventory will cover approximately 2,500 abstracting and indexing services in all subject areas. The aim has been to include all regular, generally available services covering major primary journals in a given field, which produce a maximum of 500 abstracts or 200 secondary journal references yearly. Each inventory entry comprises elements identifying the title of the service, publishers, coverage in time, publication frequency, price, volume and type of contents, indexes available, languages used, forms in which services are available (printed form, magnetic tape, etc.), complimentary services available, and subject description.

DOCUMENT ACCESS

Once a bibliographic reference has been found, the problem of obtaining access to the original document still remains. A few publishers offer copies of single articles from their own journals, upon request, and also have a depository service such as the Physics Auxiliary Publication Service of the American Institute of Physics, which supplies copies of computer printouts, tables, and other refereed material too voluminous to print in the journal. The Institute for Scientific Information has an Original Article Tear Sheet service (OATS) (30) which provides tear sheets or photocopies of articles from about 5,000 journals. New U.S. copyright laws affect the ability of some libraries to make photocopies on an "interlibrary loan" basis. A recent study sponsored by the Association of Research Libraries (31) has focussed on the solution to the interlibrary loan problem at the national level. The study examined several possible configurations for a national periodicals resource system, and recommended development of a single national center with a comprehensive collection. A similar system already exists at the British Library Lending Division (32).

An effective system for data-descriptive records will require the cooperation of primary publishers, abstracting and indexing services, and data centers to develop compatible and interchangeable methods. These groups will also have to work with libraries to ensure the availability of the primary literature.

In addition, all of this must be accomplished in a practical and economic way which will not place an insupportable financial burden on the authors, the publishers, the secondary services, the data centers, the libraries, and the other institutions involved in the chain of communication between the producer and the user of data.

REFERENCES

1. "Writing Abstracts," Z39.14-1971, American National Standards Institute, New York (1971).

2. Weil, B. H. "Standards for Writing Abstracts," *J. Amer. Soc. Inform. Sci.* **21**, pp. 351-357 (1970).

3. "Guide for the preparation of authors' abstracts for publication." In: *Guide for the preparation of scientific papers for publication*, SC/MD/5, Unesco, Paris (1968).

4. "Flagging and Tagging Data," ***CODATA Bulletin*** **19**, ICSU/AB-CODATA Joint Working Group on Tagging and Flagging, Paris, France (July 1976).

5. Lerner, R. G., R. Feinman, I. Lieblich, and S. Schiminovich "Final Report: Data-Descriptive Records in the Physical Sciences," AIP 76-1, American Institute of Physics, New York (July 1976). National Technical Information Service, Springfield, Virginia. PB-266 770/7WL.

6. Appendix 3, ***CODATA Bulletin*** **19**, loc. cit.

7. Vasta, B. M. and D. F. Zaye, "Proposed Data and Concept Tags in TOXLINE," American Chemical Society 172nd National Meeting, San Francisco, California (August/September 1976).

8. "Experiment in Data Tagging in Information Accessing Services Containing Energy-Related Data," Final Report to the National Science Foundation, DSI 75-03491, Chemical Abstracts Service, Columbus, Ohio (August 1978).

9. Kottenstette, J. P., J. E. Freeman, E. R. Staskin, and C. W. Hargrave "Accessing Numeric Data Via Flags and Tags: A Final Report on a Real World Experiment," NASA Technical Memorandum 79326 (January 1978).

10. Notes and Comments, 3, Nos. 3 and 4, Engineering Index, Inc., New York (Fall/Winter 1976).

11. Tubbs, N. "Proposed Data Referral Links Between INIS and Specialized Data Compilations," N/NF-A-41, OECD Nuclear Energy Agency (April 1976).

12. Murdock, J. W. "Current Knowledge on Numerical Data Indexing and Possible Future Developments," Informatics, Inc., Rockville, Maryland (August 1977).

13. O'Connor, J. "Data Retrieval by Text Searching." *J. Chem. Inform. Computer Sci.* **17.3**, 181 (August 1977).

14. Weinstock, M. "Abstracting and Indexing Services," in *Encyclopedia of Computer Science and Technology* **1**, 142-166 Marcel Dekker, Inc., New York (1975).

15. Bearman, T. C. "Secondary Information Services," in *Ann. Rev. Inform. Sci. Tech.* **13**, 179-208 (1978).

16. International Council of Scientific Unions, "Organization and Activities," pp. 141-144, ICSU Secretariat, Paris (October 1976).

17. "INIS International Nuclear Information System," International Atomic Energy Agency, Vienna (1970).

18. "NTIS Information Services," U.S. Department of Commerce, Springfield, Virginia (January 1976).

19. *NFAIS Newsletter*, National Federation of Abstracting and Indexing Services, Philadelphia, Pennsylvania (February 1979).

20. *International Directory of Computer and Information System Services 1974*, 3rd edition. Europa Publications, London (1974).

21. Kruzas, A. T., ed., *Encyclopedia of Information Systems and Services*, Edwards Bros., Inc., Ann Arbor, Michigan (1974) and updates.

22. Williams, M. E. and S. H. Rouse, eds., *Computer-Readable Bibliographic Data Bases - A Directory and Data Sourcebook*, American Society for Information Science, Washington, D.C. (1976).

23. Williams, M. E., L. Lannom and E. T. Dunatov *ASIDIC Survey of Information Centers Using Machine-Readable Data Bases*, Association of Information and Dissemination Centers, P.O. Box 8105, Athens, Georgia (1976).

24. *ASIDIC Newsletter*, Secretariat, Association of Information and Dissemination Centers, P.O. Box 8105, Athens, Georgia 30601.

25. *NEWSIDIC*, European Association of Scientific Information Dissemination Centres, P.O. Box 1766, The Hague, Netherlands.

26. Williams, M. E. "Use of Machine-Readable Data Bases," *Ann. Rev. Inform. Sci. Tech.* **9**, 221-284 (1974).

27. Schipma, P. B. "Generation and Use of Machine-Readable Data Bases," *Ann. Rev. Inform. Sci. Tech.* **10**, 237-271 (1975).

28. Wilde, D. U. "Generation and Use of Machine-Readable Data Bases," *Ann. Rev. Inform. Sci. Tech.* **11**, 267-298 (1976).

29. Summit, R. K. and O. Firschein "Document Retrieval Systems and Techniques," *Ann. Rev. Inform. Sci. Tech.* **9**, 285-331 (1974).

30. "Need a Scientific Article Fast?" Institute for Scientific Information, Philadelphia, Pennsylvania (1976).

31. Palmour, V. E., M. C. Bellassi, and L. M. Gray "Access to Periodical Resources: A National Plan," Association of Research Libraries, Washington, D.C. (February 1974).

32. "BLLD Publications and Other Publications Distributed by the BLLD," The British Library, Boston Spa, England (March 1976).

BIBLIOGRAPHY

B-1. *Style Manual*, American Institute of Physics, New York (1978).

B-2. *Handbook for Authors*, American Chemical Society, Washington, D.C. (1967).

B-3. *A Manual of Style*, University of Chicago Press, Chicago (1969).

B-4. *Symbols, Units and Nomenclature*, International Union of Pure and Applied Physics, Paris (1978) (Also published in *Physica* **93A**, pp. 1-60 (1978).)

B-5. *Manual of Symbols and Terminology for Physicochemical Quantities and Units*, International Union of Pure and Applied Chemistry, Pergamon, Oxford, U.K. (1979).

Data Handling for Science and Technology
S.A. Rossmassler and D.G. Watson (eds.)
North-Holland Publishing Company

COMPILATION AND EVALUATION OF DATA

D. G. Watson

ABSTRACT

This chapter identifies some of the important factors involved in the compilation and evaluation of data. These include assessment of experimental techniques, realistic estimates of errors, internal consistency, the need for easy access to the raw data, choice of "best" values. The role of the information analysis center in providing services to the user community is also discussed.

In an article on the aims of CODATA, Waddington (1) states:

> *"A half century ago it was quite feasible to plan systematically for the collection, evaluation, and publication of all or nearly all of the useful data in the scientific literature in one single coordinated effort. The publication of the International Critical Tables of Numerical Data, Physics, Chemistry and Technology... between 1926 and 1933 was the result of such an effort."*

It is certainly true that 50–70 years ago the number of data compilations was small and each compilation had fairly wide coverage. Other examples of such important publications include the *Landolt-Börnstein Tabellen, Tables Annuelles de Constantes et Données Numériques, Kaye and Laby Tables of Physical and Chemical Constants*, etc.

In the third quarter of this century, however, the amount of data in science and technology grew so rapidly that broad-coverage compilations of the above type could not be maintained. Instead most commercial publishers of data concentrated on marketing separate volumes covering specialised topics.

Another very significant development was the emergence of coordinated national plans for the organised compilation and evaluation of data in carefully chosen subject areas. In the United States, for example, the National Standard Reference Data System was established with the following objectives (2):

> *"The overall goal of the NSRDS is to provide to the technical community of the United States optimum access to critically evaluated quantitative data on the physical and chemical properties of substances and their interactions. The coverage is to be comprehensive, timely and readily accessible."*

Such national and international data activities have increased and reviews of some of them were presented at the 5th International CODATA Conference in 1976 (3–9).

According to the study published by the CODATA Task Group on Accessibility and Dissemination of Data (10), three categories of data service are necessary:

- data evaluation and compilation
- data dissemination

- data referral

The last two services will be discussed in Chapter 10 and the remainder of this Chapter will be concerned only with the first one.

Scientific data may be considered to span the vast body of knowledge which has accumulated as a result of scientific experiment, observation and calculation. One method of classifying data has been proposed (10) involving categories which are independent of the particular subject disciplines. The major categories can be summarised as:

- time-independent and time-dependent data
- location-independent and location-dependent data
- primary, derived and theoretical data
- determinable and stochastic data
- quantitative and qualitative data
- data as numerical values, graphs or models.

Another approach is to classify data according to the needs of users. This is a difficult process and should be viewed with caution since users' needs may change drastically as science and technology advance. Nevertheless the following broad categories can be recognised:

- data generated in a specific discipline and used almost exclusively by specialists in the same discipline
- data that are also used by research workers in a limited number of related disciplines
- data that are used more widely.

The critical evaluation of data as defined in a recent review (11)

> *"refers to those processes involved in assuring that published or retrieved data meet certain standards for accuracy and dependability."*

Similar definitions which stress the need to provide estimates of the errors associated with the data have been given by Kieffer (12) and Branscomb (13). Thus Kieffer states:

> *"A short definition of reliable data would be data which are presented with error bars which were chosen so that the probability of the 'true' value lying outside of these limits of error is extremely small."*

Branscomb suggests that we

> *"define critical evaluation as requiring that a meaningful quantitative statement can be made about the probable presence of systematic errors in the data. This statement must be based on a set of objective criteria for assessing the likely presence and effect of systematic errors."*

The problems of evaluation procedures have been discussed by several authors (14-16) and the views they express would suggest that the processes of data evaluation can be summarised as:

- examination and appraisal of the data, assessment of the experimental techniques with associated errors,

- re-analysis and recalculation of derived results, and

- selection of "best" values.

(Some of the special problems which are relevant to the handling of geo-data are described in Chapter 3B.)

The proper assessment of experimental techniques is a fundamental factor in the evaluation process, involving, as it does, a study of the experimental design, the way in which instrumentation has been used and the severe problem of systematic errors. Lide and Rossmassler (16) make the important point that advances in instrumentation and data analysis techniques do not of themselves necessarily lead to improvements in the reliability of published data. They cite the example of thermodynamic properties (17), where, for a set of key compounds, the number of older measurements of high reliability greatly exceeds the number of more recent reliable values.

In attempting to evaluate published data, a prerequisite is that the publication presents adequate details regarding the experiment, data reduction, choice of model, etc. The lack of sufficient details in many publications has prompted various bodies to prepare guidelines on how data should be presented in the primary literature (18-20). Such recommendations stress the need for authors to provide a proper assessment of the errors which are associated with the data. This topic (21) is treated at length in Chapter 4. In many cases it is necessary for an evaluator to have access to the raw data and Evans and Garvin (22) cite three reasons for this:

- for comparison of the same quantity over the same range of independent variables but measured by different laboratories

- for determining whether or not data sets measured in non-overlapping ranges of the independent variables can be joined to yield an expression valid over a wide range of these variables

- for use in a completely different application than was envisaged by the original investigators.

Another serious problem facing the evaluator is the presence in the primary literature of a large number of typographical errors, usually the result of inadequate proof-reading (23). This situation poses a very obvious danger for the person who is a non-specialist and uses such data directly, bypassing the evaluation process.

It is very difficult to state general rules on how published data should be evaluated but some of the principles discussed by Lide and Rossmassler (16) can be summarised here. All published data must be reduced to the same common basis before any evaluation can be made and it may not be an easy task to discover what values the authors used for fundamental constants,etc. If a set of quantities are related by well-established equations, as in thermodynamics, then the consistency of related data can be checked (24-26). Spectral data present a variety of problems and, in many instances, the quality of published collections is very low. However, in the field of infrared spectra, the Coblentz Society has developed criteria for the grading of spectra (27). Additional problems have to be faced in evaluating physical quantities that have been derived from primary spectroscopic data, such as molecular geometry, force constants, etc. Here the choice of model and the method of data analysis can introduce uncertainties which are more significant than the accuracy of the spectroscopic data. Inevitably in the process of data evaluation there is often some measure of subjectivity since evaluators must recognize the fact that certain laboratories have a reputation for careful and consistent work. Although, in theory, data evaluation should be as objective as possible there are situations where the personal experience and subjective judgement of the evaluator are important factors in the assessment process.

The end result of any evaluation process should be the presentation of "best" or "recommended" values together with quantitative estimates of the uncertainties in these values. It must be borne in mind that the "best" values derived by the critical evaluation process are only the best values at the time of review (28). Any selection of recommended values is subject to change in the light of improved measurements or evaluation. In the areas of thermodynamics and the fundamental physical constants CODATA Task Groups have made a major contribution to the compilation of key values (29-30).

Traditionally, scientists have—to varying degrees—evaluated data in their own subject field and competent individuals still find this a satisfactory procedure. Increasingly, however, various factors suggest that there is a need for a more formal and organised mechanism for data evaluation. Thus, in many fields of science the volume of published data is now so enormous that individual evaluation efforts may be almost impossible or, at best, very inefficient. Additionally there is a fast growing need for data to solve problems in multi-disciplinary projects. In this situation a scientist or technologist may have to use data which he has neither the competence nor the time to evaluate. For these reasons and others there has emerged the concept of the data evaluation center, sometimes referred to as an information analysis center, IAC (31).

Brady (32) has proposed the following definition of an information analysis center:

> *"An information analysis center is a formally structured organizational unit specifically (but not necessarily exclusively) established for the purpose of acquiring, selecting, storing, retrieving, evaluating, analyzing and synthesizing a body of information in a clearly defined specialized field or pertaining to a specified mission with the intent of compiling, digesting, repackaging or otherwise organizing and presenting pertinent information in a form most authoritative, timely and useful to a society of peers and management."*

It is clear from this definition that the total range of operations of an IAC cover the information processing chain treated in the various chapters of this book.

The core activity of the IAC, namely the evaluation process, must be carried out by professional scientists in the particular discipline. They should ideally be working close to an active research group so that they have a proper appreciation of current developments in the field but Wakelin (33) sounds an important cautionary note -

> *"... if it is too 'ivory tower,' that makes it easy for IAC staff members to consider that the specialists in their field are the primary users of their information. That would be an error, or at least an over-simplification. It is the non-expert, or the expert from another field, who has the greatest need for IAC output and services."*

As stressed in an OECD report (34) it is essential that any information system be truly active in its response to the needs of its users.

IAC's should be encouraged to publish the criteria by which they evaluate the published literature. If they can thereby gain the respect of the users and data generators then one may hope that the IAC can play a key role in promoting the better planning of experiments and reporting of the results. A classic example of this is reported by Branscomb (35) in relation to the article by Kieffer and Dunn (36) on reference data on cross-sections for ionization in electron collisions. Analysis of subsequent publications which cited this article indicated that at least one-third of these later investigations had been influenced by the critical review of Kieffer and Dunn.

The whole question of the economics of operation of information analysis centers is complex and multifaceted. It has been estimated that in high technology countries, 0.1-0.2 per cent of public funds set aside for scientific and technological research and development is allocated for data evaluation. This means that there is a huge body of data being published which will remain unevaluated, possibly for a very long time. Some experts reckon that if 1-2 per cent of the R/D budget were devoted to data evaluation then it would be possible to cope with the bulk of published data and thereby improve the efficiency of those involved in both mission-oriented and interdisciplinary projects.

Another important economic problem is the marketing of the products of IAC's. Branscomb (35) has summarised some of the important factors which limit the sales potential of such centers:

- *"Inadequate economies of scale resulting from reaching too small a fraction of the potential market;*

- *Traditionalist attitudes of the technical community towards information transfer mechanisms of new kinds, combined with*

- *A tenacious and well justified desire of scientists to reserve their dependence on information sources to these sources whose continuous availability and quality are reasonably well assured;*

- *Less than full confidence in the reliability of the information products offered, so that the user does not risk defraying a substantial cost for information with an even greater cost avoidance achieved by relying on it;*

- *Less than fully effective marketing of information analysis center products, combined with the fact that information is more efficiently and effectively wholesaled than retailed. Thus economic return depends on intermediate institutions such as libraries, which are not in a position to recover or even measure economic benefits from good information."*

REFERENCES

1. Waddington, G. "A World System of Evaluated Numerical Data for Science and Technology," *J. Chem. Doc.*, **7**, 20 (1967).

2. Brady, E. L. "The National Standard Reference Data System," *J. Chem. Doc.*, **7**, 6 (1967).

3. Sytchev, V. V., A. D. Kozlov, and L. A. Alexeeva "The National Program of Research in the Field of Numerical Data for Science and Technology." ***CODATA Bulletin*** No. **21**, 35 (1976).

4. Behrens, H. and G. Ebel "An Information System for Physics Data in the Federal Republic of Germany." ***CODATA Bulletin*** No. **21**, 38 (1976).

5. Shimanouchi, T. "Two Projects on the Evaluation and Compilation of Scientific Data in Japan." ***CODATA Bulletin*** No. **21**, 41 (1976).

6. Wysocki, A. "The Role of the UNISIST Programme in the Development of National and International Data Activities." ***CODATA Bulletin*** No. **21**, 52 (1976).

7. Eicher, L. D. and F. K. Willenbrock "A Report on the Data Concerns and Programs of the Committee on Engineering Information of the World Federation of Engineering Organizations." ***CODATA Bulletin*** No. **21**, 62 (1976).

8. Huber, W. "Data Banks and EURONET." ***CODATA Bulletin*** No. **21**, 64 (1976).

9. Lorenz, A. "The IAEA Nuclear Data Center. Its Role in the International Scientific Community." ***CODATA Bulletin*** No. **21**, 67 (1976).

10. "Study on the Problems of Accessibility and Dissemination of Data for Science and Technology." ***CODATA Bulletin*** No. **16**, (1975).

11. Luedke, J. A., Jr., G. J. Kovacs, and J. B. Fried "Numeric Data Bases and Systems," *Ann. Rev. Infor. Sci. Tech.*, Vol. **12** Knowledge Industry Publications, Inc., White Plains, New York (1977).

12. Kieffer, L. J. "The Reliability of Property Data, or, Whose Guess Shall We Use?" *J. Chem. Doc.*, **9**, 167 (1969).

13. Branscomb, L. M. "Is the Literature Worth Retrieving?" *Sci. Res.*, **3**, 49 (1968).

14. Compton, W. D. "Our Objective: Systems with Critically Evaluated Data," *Bull. Amer. Soc. Info. Sci.* **1**(7):3,37 (1975).

15. Lide, D. R. Jr. "The National Standard Reference Data System," *Bull. Amer. Soc. Info. Sci.* **1**(7):10,34 (1975).

16. Lide, D. R., Jr., and S. A. Rossmassler "Status Report on Critical Compilation of Physical Chemical Data," *Ann. Rev. Phys. Chem.* **24**, 135 (1973).

17. "Tentative Set of Key Values for Thermodynamics-Part I." ***CODATA Bulletin*** No. **2** (1970) see also ref. 29.

18. "Guide for the Presentation in the Primary Literature of Numerical Data Derived from Experiments." ***CODATA Bulletin*** No. **9** (1973).

19. "The Presentation of Chemical Kinetics Data in the Primary Literature." ***CODATA Bulletin*** No. **13** (1974).

20. "Recommendations for Measurement and Presentation of Biochemical Equilibrium Data." ***CODATA Bulletin*** No. **20** (1976).

21. Tukey, J. W. "Modern Statistical Techniques in Data Analysis." ***CODATA Bulletin*** No. **21**, 8 (1976).

22. Evans, W. H. and D. Garvin "The Evaluator versus the Chemical Literature." *J. Chem. Doc.*, **10**, 147 (1970).

23. Evans, W. H. "Effects of Errors in the Chemical Literature on the Compilation of Critically Evaluated Data," *J. Chem. Doc.*, **6**, 135 (1966).

24. Bondi, A. "On Error Prevention," *J. Chem. Doc.*, **6**, 137 (1966).

25. Bondi, A. "Design of Papers for Error Minimization," *J. Chem. Doc.*, **9**, 7 (1969).

26. Guest, M. F., J. B. Pedley and M. Horn "Analysis by Computer of Thermochemical Data on Boron Compounds," *J. Chem. Thermodyn.* **1**, 345 (1969).

27. The Coblentz Society Board of Managers. "Specifications for Evaluation of Infrared Reference Spectra," *Anal. Chem.* **38**, 27A (1966).

28. Goudsmit, S. A. "Is the Literature Worth Retrieving?" *Phys. Today* **19**, 9 (1966).

29. CODATA Recommended "Key Values for Thermodynamics." ***CODATA Bulletin*** No. **28** (1978).

30. "Recommended Consistent Values of the Fundamental Physical Constants." ***CODATA Bulletin*** No. **11** (1973).

31. "The Management of Information Analysis Centers." Proceedings of a Forum sponsored by the COSATI Panel on Information Analysis Centers. COSATI Report No. 72-1 (1972) (available from NTIS, U.S. Dept. of Commerce, 5285 Port Royal Road, Springfield, Virginia 22151, USA).

32. Brady, E. L. "Objectives of the Forum and the Activities of COSATI Panel No. 6." Proceedings of the Forum of Federally Supported Information Analysis Centers. COSATI PB 177051 (1967). (Available from NTIS, U.S. Dept. of Commerce, 5285 Port Royal Rd., Springfield, Virginia 22151, USA).

33. Wakelin, J. H., Jr. "The National Stake in Better Technical Information." See reference 31, p. 105.

34. "Information for a Changing Society. Some Policy Considerations." Organisation for Economic Cooperation and Development, Paris (1971).

35. Branscomb, L. M. "Information Analysis Centers: The Challenge of Being Needed." Reference 31, p. 3.

36. Kieffer, L. and G. Dunn "Electron Impact Ionization Cross-Section Data for Atoms, Atomic Ions and Diatomic Molecules; I. Experimental Data," *Rev. Mod. Phys.* **38**, 1 (1966).

BIBLIOGRAPHY

B-1. Lide, D. R. Jr., and M. A. Paul, ed. *Critical Evaluation of Chemical and Physical Structural Data.* Nat. Acad. of Sciences, Washington, D.C., USA, (1974).

B-2. Weisman, H. M. "Needs of American Chemical Society Members for Property Data," *J. Chem. Doc.*, **7**, 9 (1967).

B-3. Weisman, H. M. "A Survey on the Use of National Standard Reference Data System Publications," *J. Chem. Doc.*, **12**, 211 (1972).

B-4. Mountstephens, B., A. Osborn, and M. Slater *Quantitative Data in Science and Technology.* Aslib Occasional Publication No. **7**, (1971).

B-5. Slater, M., A. Osborn, and A. Presanis *Data and the Chemist.* Aslib Occasional Publication No. **10**, (1972).

B-6. Darby, R. L. "Information Analysis Centres as a Source for Information and Data." *Special Libraries* **59**, 91 (1968).

B-7. Martyn, J. *Notes on the Operation of Specialised Information Centres.* Aslib Occasional Publication No. **5**, (1970).

B-8. "UNISIST Guidelines for Establishing and Operating Information Analysis Centers." Prepared with the collaboration of F. Kertesz. Paris, Unesco unpublished report.

B-9. Proceedings of the Forum of Federally Supported Information Analysis Centers. COSATI PB 177051, (1967). (available from NTIS, U.S. Dept of Commerce, 5285 Port Royal Rd., Springfield, Virginia 22151, USA).

B-10. Kehiaian, H. V., "Problemes de collecte et de compilation de données," Communication au colloque CODATA-France "Utilisation de données--Banques de données," Paris, (September 1977).

B-11. "Miniconfrontation on Information Analysis Centers." OECD, (1970).

B-12. Rossmassler, S. A. "Public/Private Cooperation in Planning and Developing Reference Data Programs." *J. Chem. Doc.*, **13**, 65 (1973).

Data Handling for Science and Technology
S.A. Rossmassler and D.G. Watson (eds.)
North-Holland Publishing Company

STANDARDS AND GUIDELINES FOR DATA

P. W. Berger and J. C. Tucker

ABSTRACT

Measurements are made in laboratories to establish hypotheses, control industrial processes, aid in health care, and set standards. In order to be most useful, these measurements must be precise and must also be compatible with one another. To provide precise numerical values, standard reference data systems produce critically evaluated compilations of numerical values of physical and chemical properties. International attention to needs for standard reference data is provided by CODATA; there are also several national reference data programs. Standard reference materials with accurately assessed properties for comparison of instruments and methods are provided by several national programs listed by the International Union of Pure and Applied Chemistry. These reference materials must be used in conjunction with standard reference methods in laboratories to assure compatibility of results. Engineering and product standards are written by a number of national and international organizations to assure the quality and compatibility of industrial output. Standardization of bibliographic records and format structures for such records increases the effectiveness of transfer of information and data. Finally, major international attention has been focussed on standardization of information processing, exemplified by the Patent Cooperation Treaty, the International Organization for Standardization, the International Electrotechnical Commission, plus the World Meteorological Organization and many other United Nations Agencies.

MEASUREMENT SYSTEMS

Scientific experiments produce data which are published and evaluated, as outlined in the preceding chapters. In order to apply the results to further research or application, not only is critical evaluation of the laboratory procedure and data required, but the data must be related to universally acceptable units of measure so as to be usable in other laboratories. *"It is the obligation of each investigator to place his observations and measurements on a solid foundation by appropriate linkage with the fundamental units of measurement involved."* (1) It is at this stage that standards become important. Agreement must first be reached on the units to be used in measurements, since units of measure developed from local needs and materials may not be comparable nationally or internationally. The Système International d'Unités (SI) for units of measure, with the symbols in which they are expressed (2), is based on the metric system and has been widely adopted as an international standard for units of measure because of its coherent and rational character.

Once the units have been defined, it becomes necessary to assure that measurements made in one laboratory are the same as measurements made in another laboratory on the same quantity, substance or the like, i.e., the laboratories must be "compatible." Both instruments and methods used in the laboratory must be standardized. Compatibility among laboratories may be achieved either by calibration--having one laboratory, designated as the "standard," adjust the instruments of all other laboratories within a group--or by having the "standard" laboratory provide a measurable entity the other laboratories can then measure, adjusting their instruments and methods to achieve the tagged results. The latter method has the advantage of verifying the procedures as well as the instruments of the coordinating laboratories. The process of matching initial measurements to a compatible system is described in detail by Huntoon (3). Many individual countries support one or more national laboratories which have some degree of responsibility for measurement standardization. International coordination and compatibility are effected through the work of the International Bureau of Weights and Measures (BIPM) (4). In the United States and Canada, the National Conference of Standards Laboratories brings together organizations which are interested in measurement standards and calibration facilities (5).

The measurable entity to be supplied by the standard laboratory may be an operational definition for realizing a fundamental measurement unit, standard reference data, standard reference materials, or a standard transfer instrument which has been calibrated by the standard laboratory. A standard transfer instrument is used to make measurements which are compared to the measurements of the laboratory's own instrument, which is then adjusted to produce the same results as the standard instrument. This method of developing compatibility will not be discussed further. In the following discussion, the definition of the fundamental physical constants is treated as a special case of standard reference data.

STANDARD REFERENCE DATA

If enough is known about a material or physical property so that it can be reproduced in any laboratory when the correct methods are followed, then it is sufficient for standardization purposes to publish standard values of data for the properties to be measured. Beginning about 1963, there was recognition among the developed nations of the need for an international effort to compile, to evaluate, and to make available such critical values of well-defined physical and chemical properties (discussed in Chapter 7). National programs to evaluate numerical reference data were mounted in the United States, the Union of Soviet Socialist Republics, the United Kingdom, Japan, France, and Germany, and other nations have since joined (6,7). These programs are listed in Appendix I. At the same time, these nations recognized the need for an international coordinating effort . Accordingly, the International Council of Scientific Unions, ICSU, established a Committee on Data for Science and Technology, CODATA (8). The development of CODATA is noted in Chapter 1. Evaluation work is coordinated by ten CODATA Task Groups:

- Accessibility and Dissemination of Data
- Chemical Kinetics
- Computer Use
- Data for the Chemical Industry
- Fundamental Constants
- Internationalization and Systematization of Thermodynamic Tables
- Key Values for Thermodynamics

- Space and Time Dependent Data
- Transport Properties for Solids
- International Training Courses on the Handling of Experimental Data

and by the Advisory Panels:

- Bioscience Advisory Panel
- Industrial Data Advisory Panel
- Subcommittee on Geosciences

The chief means of communication of the results of CODATA activities are the ***CODATA Newsletter***, the ***CODATA Bulletin***, and the ***International Compendium of Numerical Data Projects***. International conferences are held on the generation, compilation, evaluation and dissemination of data for science and technology (9,10).

A major adjustment in the precision of the fundamental constants was published by the National Standard Reference Data System in the U.S. and adopted by CODATA (11) in 1973. Numerous compilations of specific property data have been published by the various national programs or by the Task Groups of CODATA. It has been proposed that CODATA produce a General Index of all such compilations and act as a clearinghouse to provide access to them (mentioned in Chapter 10). Guidelines for the publication of data in order to provide sufficient information for critical evaluation are discussed in Chapter 7 and cited also in Chapter 5.

STANDARD REFERENCE MATERIALS

In many cases, however, the realization of physical measurements from base units or the specification of a property of a material is difficult or impossible, without the material itself. *"A Standard Reference Material (SRM) ....is a material (not necessarily a pure substance) having given properties with numerically assessed values, within given tolerances, certified by an appropriate technical body."* (12) Kienitz lists the conditions under which a Standard Reference Material is required (13). Like Standard Reference Data, a Standard Reference Material must be measured and evaluated, and then its properties may be certified to the user (14). However, the substance to be used as a reference must be supplied to the laboratory using it, because it cannot be reliably reproduced and tested against published data alone. Reference materials serve as analog standards by which other materials may be evaluated. A Standard Reference Material can also provide a point of referral for reporting data in the literature as a new scientific discipline develops (15). A group of national programs to produce reference materials has been developed (16,17,18).

The use of Standard Reference Materials emphasizes the necessity for standard methods and analysis of errors in laboratory measurements, even where the laboratories themselves are not part of the standard setting system. As reference materials were used in clinical chemistry, it was discovered that no accurate measurement system existed in hospital clinical laboratories. The National Bureau of Standards in the United States is participating in programs to develop laboratory certification programs, to assure that critically evaluated data can be meaningfully used (19).

ENGINEERING STANDARDS AND PRODUCT STANDARDS

Standards play a role in the control of technology as well as in scientific investigation. In fact, many people are referring only to this area of industrial and product standards when they speak of "standardization" (20,21,22). The aims of standardization at this applied level are simplification of the growing variety of products and procedures in human life; communication of information; overall economy of production and consumption; safety, health, and protection of life; protection of consumer and community interests; and the elimination of trade barriers (23). An engineering or product standard must define the property to be measured, the method of measuring it, and the amount of variability which can be tolerated. Thus, standards of this type depend on measurement of data in comparable units from manufacturer to user. Industrial standards cover specification of physical properties, chemical composition, performance, and manufacturing methods; nomenclature; dimensional standards; and testing methods for determining compliance with the specifications. Compliance may be attested by certification marking, informative labeling, pre-shipment inspection of exports, or statistical quality control.

In many countries, standards are incorporated into the laws and regulations of government. At the same time, in some countries, voluntary standards are also established by agreement among interested parties--producers, users, retailers or middlemen, and technologists concerned with the state of the art in manufacture and measurability. In cases where the safety of human life and property are concerned, mandatory standards may be incorporated into laws and regulations.

Standards may be included in legislation by quoting the standard verbatim, by referring to the standard by number, title, and date of publication or by specifying compliance with the law in terms of the standards issued by specified authorities. The last method leaves the detailed specification of the standard, and its periodic updating as required, in the hands of standardizing organizations rather than requiring the development of legislative expertise and continual revision of legislation. The exhaustive description of the process and methods of standardization by Verman (24) includes a list of national standards-writing organizations and discussions of the role of regional and international standards-writing organizations. The Japanese Standards Association provides a case history of the development of standardization in one country (25). Some of the more important regional and international efforts are discussed in the remainder of this chapter.

STANDARDIZATION AND THE TRANSFER OF INFORMATION AND DATA

The transfer of scientific and technical data between countries or within a given country is essential for the economic health and social prosperity of nations. Indeed, the speed of information and data transfers often determines the rates of scientific advancement and achievement. Whether or not the crucial interrelationships between multiple scientific disciplines or the various elements of science and technology can be correctly interpreted and accurately applied depends, by and large, on the degree to which information and data about these interrelationships have been codified and standardized.

UNISIST early recognized the importance of the interchangeability of scientific information and data. In a 1971 feasibility study for a World Science Information System, UNISIST noted that (26):

"Scientific information embodies the heritage of man's scientific knowledge. It constitutes an essential resource for the work of scientists. It is a cumulative

resource; knowledge builds on knowledge as new findings are reported. It is an international resource, built painstakingly by scientists of all countries without regard to race, language, color, religion, or political persuasion. As it is built internationally so it is used internationally. Scientists who are its builders and users ask only that each other's contributions be verifiable; it is, therefore, not only a resource; it is a means through which the world's scientists maintain their discipline. It is a medium for the education of future scientists and a principal reservoir of concepts and data to be drawn on for application to economic and technological development programmes."

The establishment of national and international standardizing organizations and document repositories has done much to improve the intelligibility, accessibility and transferability of scientific and technical data. However, much remains to be accomplished before it can be stated that there exists a world-wide system which permits *"the unimpeded exchange of...scientific information and data among all countries...(and) hospitality to...(diverse) fields of science and technology...(as well as to the diverse) languages used for the international exchange of scientific information."* (27) What follows is a brief description of the work in this area by selected international programs, projects, and organizations.

INTERNATIONAL BIBLIOGRAPHIC STANDARDIZATION EFFORTS

Librarians and information scientists recognized the importance and necessity of bibliographic standardization long before the advent of automated information storage and retrieval systems. For example, the enunciation of the Paris Principles and the Principles of Universal Bibliographic Control in 1961 specifically acknowledged the need for such standards. Machine systems require extra measures to assure citation conformity if truly efficient machine processing is to be attained. This is so because while variation in the formatting of data elements may not disturb human interpretation of technical material, such exceptions guarantee "faulty" machine processing and therefore, inadequate or incomplete data and information retrieval. Perhaps the greatest single impetus to the standardization of automated bibliographic records in libraries was the development of the MARC (Machine Readable Cataloging) format by the Library of Congress in the 1960's (28-33).

In 1974 UNISIST published a Manual (33) designed to *"define for most types of scientific and technical literature...a set of data elements which will constitute an adequate bibliographic citation...The (working) group's purpose has been to define a minimum set of data elements which could be agreed upon by abstracting and indexing services, to facilitate the exchange of information between services, and to enable them to present their computer-based products to the user in a more compatible and therefore easily usable form."* The scope of the manual is limited to *"defining...data elements as they should appear...for exchange purposes between two or more computer-based systems."* The manual relies heavily on the work of the International Organization for Standardization (ISO), and the bibliographic exchange format described specifically implements ISO Standard 2709 (30). Bibliographic data elements and data fields for serials, books, reports, theses and dissertations, patent documents, and conference publications are carefully defined and described.

The International Federation of Library Associations and Institutions has compiled a most useful report (35) on standardization activities within libraries and with regard to national bibliographies, to facilitate

- *"reaching agreement on the minimum standards and acceptable practices for the coverage, content, and form of national bibliographic records, taking into account the requirements for their international exchange;*

- *reaching agreement also on acceptable guidelines for the presentation, arrangement, and frequency of the printed national bibliography;*

- *discussing and making proposals for the sharing of resources to assist countries to realize national bibliographic control."*

The history of IFLA's role in the development of the ISBD (International Standard Bibliographic Description), as well as IFLA's relationship to ISO is documented in a 1977 issue of the *Library of Congress Information Bulletin*(36). Finally, the development and standardization of unique indentifiers for serial publications, with special emphasis on the CODEN and ISSN systems (the latter developed under ISDS), is comprehensively documented in a bibliography compiled and annotated by Groot (37).

PATENT COOPERATION TREATY

In September of 1966, the Director of the United International Bureau for the Protection of Intellectual Property (BIRPI) was asked to undertake a study to reduce the filing and processing procedures then required for separate patent applications on the same invention in a number of countries (38). The results of this investigation culminated on June 19, 1970, when the *Patent Cooperation Treaty* was signed by twenty nations (38). The general purpose of the Treaty is *"to simplify the filing of patent applications on the same invention in different countries."*(38) Chapter I stipulates that applicants will file a single international application in an agreed language and in a standard format (39). Special attention is given to the needs of the developing nations. The preamble notes that the signatories wish *"to foster and accelerate the economic development of developing countries...by providing easily accessible information on the availability of technological solutions applicable to their special needs and by facilitating access to the ever expanding volume of modern technology..."* (40). Further, Article 51 of Chapter IV stipulates the formulation of a Committee *"to organize and supervise technical assistance for...developing countries...(to include)...the training of specialists, the loaning of experts, and the supply of equipment both for demonstration and for operational purposes."* (41)

A relatively complete legislative history of the Patent Cooperation Treaty is set out in Volume 2 of *U.S. Code, Congressional and Administrative News*, 94th Congress, 1st Session, 1975, pp. 1220-1244 (42). [In the United States, the *Patent Cooperation Treaty* was ratified in the Congress in 1975. See Public Law 94-131, 89 Stat. 685 (43).]

Post 1970 ratifications and other implementations of the PCT are documented in the monthly issues of *Industrial Property*, the official publication of the World Intellectual Property Organization (WIPO) (44). (WIPO is the successor organization to BIRPI.) In addition, *Industrial Property* reports the activities, official sessions, and pronouncements of organizations such as the Organization for Cooperation in Information Retrieval Among Patent Offices (ICIREPAT). For example, in 1976, the Plenary Committee (PLC) of ICIREPAT surveyed the patent documentation and patent information retrieval needs of the developing countries, especially the member states of the African Intellectual Property Organization (OAPI) (45). Also during 1976, the ICIREPAT's Technical Committee for Standardization (TCST) discussed the creation of international patent data codes (INID), country codes, and standard patent document formats (46). (A somewhat dated overview of the international data problems associated with patents and trademarks in general can be found in a series of lectures published by WIPO in 1971 (47).

INTERNATIONAL ORGANIZATION FOR STANDARDIZATION (ISO)

Central to all international standardization activity is the work of the International Organization for Standardization (ISO), the successor agency to the International Federation of the National Standardizing Associations (ISA). The ISO Secretariat is located in Geneva, Switzerland, and in 1976 the ISO membership included 63 organizations *"most representative of standardization in their respective countries. Only one body in each country may be admitted to membership"* (48). In addition to the member bodies, standards organizations in countries which have not yet developed national standards activity may join ISO as correspondent members. In 1976, 18 agencies and activities were correspondent members of ISO. ISO publishes a directory of its membership which lists (by country) the member organization's name, legal status, budget, and staff size, describes its sources of revenue, membership, and tabulates its standardization responsibilities and publications. A brief history of the ISO member organizations is also included, as is a statement of the status of the country's national standards (48). ISO publishes an annual catalog of its publications (49) which lists the names and current addresses of the member bodies. ISO member bodies act as sales agents for all ISO publications in their own countries (49). More than 70% of the ISO member bodies are either governmental institutions or organizations incorporated by public law. Member bodies enjoy full voting rights, participate on any technical Committee (in 1975, there were 160 Technical Committees), and are eligible for ISO Council membership.

Correspondent members do not participate in ISO's activities but are kept fully informed about the work of the Technical Committees. For a full description of the missions and organizations of the various ISO Technical Committees and their subcomponents, the reader is referred to ISO's annual *Memento* (50).

INTERNATIONAL ELECTROTECHNICAL COMMISSION (CEI)

The CEI was founded in 1906 in response to a resolution (50) passed by the 1904 International Electrotechnical Congress:

> *"That steps be taken to secure the cooperation of the technical societies of the world by the appointment of a representative Commission to consider the question of the standardization of the Nomenclature and the Ratings of Electrical Apparatus and Machinery."*

In 1947, CEI became the Electrical Division of ISO. Its organization, structure and committee memberships and publications are detailed in participant handbooks. Reference (51) cites the IEC Handbook for U.S. participants.

UNITED NATIONS STANDARDIZATION ACTIVITY

A total of 16 United Nations agencies monitor, publish, or coordinate international standards activities. Each is concerned with data standardization to the extent that meetings and publications are devoted to the standardization of technical vocabularies, glossaries, and technical symbols as well as the formatting and presentation of information and technical data.

For example, the 137 States and Territories who are members of the Organisation Mondiale pour la Métrologie (OMM), have as two of their objectives (52)

- *"to promote the establishment and maintenance of systems for the rapid exchange of metrological information.*

- *"to promote standardization of metrological observations and insure the uniform publication of observations and statistics."*

and their publications do indeed reflect their concerns in these areas. The OMM membership meets every fourth year, and its Regional Associates include representatives from Africa, Asia, South, North and Central America, the Southwest Pacific, and Europe. Information on other U.N. Agencies concerned with international standardization can be obtained by consulting the U.N. *Yearbook*, which is updated annually (53) or by contacting the appropriate national standards organization which is that nation's official member to international standards associations. In the United States, the appropriate organization is the American National Standards Institute (ANSI) (54).

SELECTED MAJOR COORDINATING REGIONAL STANDARDS BODIES

- Arab Organization for Standardization and Metrology—established as a specialized agency of the League of Arab States with headquarters in Cairo, Egypt.

- Asian Standards Advisory Committee (ASAC)—a U.N. sponsored body, established in 1966 to coordinate the work among existing national standard bodies and to facilitate the establishment of new ones.

- European Committee for Standardization (CES)—established in 1961. Central secretariat located in Paris. Created to unify and distribute standards in fields of special interest to the governments of Europe. Membership consists of the national standards organizations of the European Communities (EEC) and the European Free Trade Association (EFTA), each country having only one member.

- Pan American Standards Commission (COPANT)—founded in 1961. Secretariat—Buenos Aires, Argentina. A coordinating body, comprised of the national standards agencies of twelve Latin American countries.

- Pacific Area Standards Congress (PASC)—formed in 1973 to facilitate participation by the Pacific area states in ISO and IEC.

REFERENCES

1. Rossini, F. D. *Fundamental Measures and Constants for Science and Technology.* Cleveland, CRC Press (1974). p. 1.

2. ISO/R 1000-1969 *Rules for the Use of Units of the International System of Units and a Selection of the Decimal Multiples and Submultiples of the SI Units.* International Organization for Standardization, Geneva (1969). Essentially equivalent national documents exist in many languages. For guidance in obtaining such documents, readers are referred to their CODATA National Committees described in Chapter 1.

3. Huntoon, R. D. "Standard reference materials and meaningful measurements; an overview." In ref. 16, p. 4-56.

4. *Le Bureau International des Poids et Mesures,* 1875-1975. Sèvres, France, BIPM, 1975. *The International Bureau of Weights and Measures, 1875-1975:* translation edited by Chester H. Page and Paul Vigoureux, U.S. National Bureau of Standards, Washington, D.C. (1975). (NBS Special Publication 420).

5. Activities of this organization are reported in its *NCSL Newsletter,* available from NCSL Secretariat, National Bureau of Standards, Boulder, Colorado 80303.

6. International Symposium on Numerical Reference Data for Science and Technology, Warsaw, 1969. *Proceedings* Ossolineum, Warsaw (1973).

7. Proceedings of the Plenary Sessions, Fifth International CODATA Conference on Generation, Compilation, Evaluation, and Dissemination of Data for Science and Technology, Boulder, Colorado, USA, June 28-July 1, 1976. ***CODATA Bulletin*, 21,** (October 1976).

8. Rossini, F. D. "The ICSU Committee on Data for Science and Technology (CODATA)," *J. Chem. Doc.* **10**, 261-264 (1970).

9. International CODATA Conference, 4th, Tsakhcadzor, 1974. *Generation, Compilation, Evaluation, and Dissemination of Data for Science and Technology.* Pergamon Press, New York (1975).

10. International CODATA Conference, 5th, Boulder, Colo., 1976. *Generation, Compilation, Evaluation, and Dissemination of Data for Science and Technology* Pergamon Press, New York (1977).

11. Cohen, E. R. and B. N. Taylor "The 1973 Least-Squares Adjustment of the Fundamental Constants." *J. Phys. Chem. Ref. Data,* **2**, 663-734 (1973).

12. Milazzo, G. "Selection criteria of a material as standard reference material and steps for certification." In ref. 16, p. 127.

13. Kienitz, J. "Calibration and test materials for physicochemical measurements (Report on the work of IUPAC Subcommission I.4.1)." In ref. 16, p. 118-126.

14. Milazzo, G. "Selection criteria of a material as standard reference material and steps for certification." In ref. 16, p. 127-145.

15. Mears, T. W. "SRM needs and measurement problems in science-chemical properties." In ref. 16, p. 395-410.

16. Materials Research Symposium, 6th, National Bureau of Standards, 1973. *Standard Reference Materials and Meaningful Measurements.* U.S. National Bureau of Standards, Washington, D.C. (1975). (NBS Special Publication 408)

17. Cali, J. P. and C. L. Stanley "Measurement Compatibility and Standard Reference Materials." In *Annu. Rev. Mater. Sci.* **5**, 335-339 (1975).

18. Cali, J. P. "The NBS Standard Reference Materials Program: An Update." *Anal. Chem.* **48**, 802A-818A (1976).

19. Gnaedinger, J. P. "Are we ready for standards on laboratory competence?" *Materials Research and Standards*, **12**, No. 11, 8-13 (1972).

20. Coonley, J. and P. G. Agnew, *The Role of Standardization in the System of Free Enterprise.* American National Standards Association, New York (1941).

21. Melnitsky, B. *Profiting from Industrial Standardization.* Conover-Mast, New York (1953).

22. Reck, Dickson, ed. *National Standards in a Modern Economy.* Harper, New York (1956).

23. Sanders, T. R. B., ed. *The Aims and Principles of Standardization.* Geneva, International Organization for Standardization (1972). p. 5-11.

24. Verman, L. C. *Standardization: A New Discipline.* Archon Books, Hamden, Conn. (1973).

25. Japanese Standards Association. *Progress of Industrial Standardization in Japan: 50 Years of Japanese Standards.* Japanese Standards Association, Tokyo (1973).

26. UNISIST. *Synopsis of the Feasibility Study on a World Science Information System.* Unesco, Paris (1971). p. v-vi

27. Ibid., p. vi-vii.

28. Avram, H. D. *MARC, Its History and Implications.* U.S. Library of Congress, Washington, D.C. (1975). 49 p.

29. American National Standards Institute. *American National Standard Format for Bibliographic Information Interchange on Magnetic Tape.* New York (1971). 34 p. ANSI Z39.2-1971.

30. International Organization for Standardization. *Documentation - Format for Bibliographic Information Interchange on Magnetic Tape.* Geneva (1973). 4 p. ISO 2709-1973 (E).

31. Rather, L. J. and P. J. De La Garza "Getting It All Together: International Cataloging Cooperation and Networks." *J. Libr. Autom.* **10**, 163-169 (June 1977).

32. International Federation of Library Associations (comp.) *Standardization*

Activities of Concern to Libraries and National Bibliographies: An Outline of Current Practices, Projects and Publications. IFLA Committee on Cataloguing, London (1976). p. 18.

33. *Reference Manual for Machine-Readable Bibliographic Descriptions.* Paris (1974). p. 5.

34. See Reference 30.

35. International Federation of Library Associations (Comp.) *Standardization Activities of Concern to Libraries and National Bibliographies: An Outline of Current Practices, Projects and Publications.* IFLA Committee on Cataloguing, London (1976). p. 163-169, p. vii.

36. Anderson, D. "A Rejoinder: IFLA's Role in Standardizing Bibliographic Practices - The ISBD Program." *Library Congress Information Bull.* **36** (August 26, 1977) p. 600-604.

37. Groot, E. H. "Unique Identifiers for Serials: An Annotated, Comprehensive Bibliography." *The Serials Librarian,* **1**, 51-75 (Fall 1976).

38. *The Patent Cooperation Treaty.* U.S. Patent Office, Washington, D.C. (1971). pp. 1-3.

39. Ibid., p. 12.

40. Ibid., p. 7.

41. Ibid., p. 41.

42. *U.S. Code, Congressional and Administrative News, 94th Congress, 1st Session, 1975.* West Publishing Co., St. Paul, Minnesota (1976). Two vols.

43. Public Law 94-131 (S. 24); November 14, 1975. *Patent Cooperation Treaty.* (89 Stat. 685). In *U.S. Code, Congressional and Administrative News, 94th Congress, 1st Session, 1975,* Vol. 1.

44. *Industrial Property, Monthly Review of the World Intellectual Property Organization.* Geneva (1961-).

45. Ibid., (March 1977). p. 74.

46. Ibid., p. 74-75.

47. *Current Trends in the Field of Intellectual Property.* World Intellectual Property Organization, Geneva (1971). 399 p.

48. International Organization for Standardization. *ISO Member Bodies.* Geneva (1974). 60 p.

49. International Organization for Standardization. *ISO Catalogue, 1976.* Geneva (1976). 177 p.

50. International Organization for Standardization. *ISO Memento.* Geneva. Issued annually.

51. *IEC Participation: Handbook for Members of U.S. National Committees.* American National Standards Institute, New York (1977). (looseleaf)

52. World Metrological Organization. *Catalogue of Publications.* Geneva (1973). 106 p.

53. United Nations. *Yearbook.* Columbia University Press, New York (1946/47). Issued annually.

54. *The ABC's of International Standardization.* American National Standards Institute, New York (1975).

BIBLIOGRAPHY

B-1. Anderson, D. "A Rejoinder: IFLA's Role in Standardizing Bibliographic Practices - the ISBD Program." *Library of Congress Inform. Bull.*, **36**, (August 26, 1977). pp. 600-604.

B-2. *ANSI Progress Report, 1977*. American National Standards Institute, New York (1977). 40 p. Describes organization, activities and composition of the Councils and Boards which are the deliberating and coordinating bodies of the Council. Includes a list of the members.

B-3. Astin, A. V. "Report on the Symposium on an International Standard Reference Materials Program." *Metrologia* **6**, 33-34 (1970).

B-4. Avedon, D. M. "Standards: 1977 ISO Report." *J. Microgr.* **10**, 343-346. (1977). Reports on 16th meeting of ISO's Technical Committee 46/Subcommittee 1, on Documentary Reproduction.

B-5. Avram, H. D. *MARC: Its History and Implications*. U.S. Library of Congress, Washington, D.C. (1975). 49 p. Pages 21-24 describe the influence of the MARC format on standardization efforts in the U.S. and internationally. Pages 41-44 contain bibliographic references to article, symposia, etc. regarding adaptations of MARC by other countries and the development of SUPERMARC (now UNIMARC).

B-6. Barry, K. M., et al., "Recommended Standards for Digital Tape Formats, *Geophysics*, **40**, 344-352. (No. 2, 2975). The use of minicomputers in digital field recording has created a demand for demultiplexed formats for seismic data. In addition, the seismic industry has experienced a growing need to standardize an acceptable data exchange format. The Seg Y format is proposed to accommodate both these needs.

B-7. Bundesanstalt für Materialprüfung (BAM). *Jahresbericht*, Berlin (1972).

B-8. Bureau of Analysed Samples. *British chemical standards and spectrographic standards*. Newham Hall, Middlesborough, Teesside, England (1973). (List number 448).

B-9. Cali, J. P. "Needs for standard reference materials for calibration and quality control." In *Metrology and Standardization in Less Developed Countries: The Role of a National Capability for Industrializing Economies*. U.S. National Bureau of Standards, Washington, D.C. (1971). (NBS Special Publication 359) p. 95-105.

B-10. Cali, J. P. "The NBS Standard Reference Materials Program: An Update," *Anal. Chem.* **48**, 802A-818A (1976).

B-11. Cali, J. P. "Problems of standardization in clinical chemistry," *Bulletin of the World Health Organization* **48**, 721-726 (1973).

B-12. Cali, J. P. et al., *The Role of Standard Reference Materials in Measurement Systems*. U.S. National Bureau of Standards, Washington, D.C. (1975). (NBS Monograph 148).

B-13. Cali, J. P. and C. L. Stanley "Measurement compatibility and standard reference materials," *Annu. Rev. Mater. Sci.* **5**, 329-343 (1975).

B-14. Chumas, S. J. *Directory of United States Standardization Activities*, U.S. National Bureau of Standards, Washington, D.C. (1975). (NBS Special Publication 417). Describes activities in the fields, products, and services in which 580 U.S. organizations specialize.

B-15. Chumas, S. J. *Index of International Standards.* U.S. National Bureau of Standards, Washington, D.C. (1974). 222 p. (NBS Special Publication 390). An index to the standards of the International Organization for Standards (ISO), The International Electrotechnical Commission (IEC), The International Commission on Rules for the Approval of Electrical Equipment (CEE), The International Special Committee on Radio Interference (CISPR) and The International Organization of Legal Metrology (OIML). Includes a list of organizational names and addresses and a list of organizations by country from which the standards of other countries and international standards may be obtained.

B-16. Chumas, S. J. *Tabulation of Voluntary Standards and Certification Programs for Consumer Products.* U.S. National Bureau of Standards, Washington, D.C. (1973). (NBS Technical Note 762). National, industrial and international standards which deal primarily with either safety or performance or both aspects of products. Also provides available information on certification programs and standards under development, and standards industrial classification numbers for the products.

B-17. Cochrane, R. C. *Measures for Progress.* U.S. National Bureau of Standards, Washington, D.C. (1966). A history of the United States National Bureau of Standards.

B-18. Cohen, E. R. and B. N. Taylor "The 1973 least-squares adjustment of the fundamental constants," *J. Phys. Chem. Ref. Data,* **2**, 663-734 (1973).

B-19. Commission of the European Communities. *Community Survey on Standard Reference Materials.* Luxembourg (1973).

B-20. Commission of the European Communities, Community Reference Bureau. *Reference Materials.* Office of Official Publications of the European Communities, Luxembourg (1973).

B-21. *Critical surveys of data sources.* U.S. National Bureau of Standards, Washington, D. C. (1974-). (NBS Special Publication 396). A series of guides for groups which select and compile data on the properties of commercial materials, particularly mechanical properties.

1. Gavert, R. B., R. L. Moore, and J. H. Westbrook *Mechanical Properties of Metals* (1974).
2. Johnson, D. M. and J. F. Lynch *Ceramics* (1975).
3. Diegle, R. B. and W. K. Boyd *Corrosion of Metals* (1976).
4. M. J. Carr et al. *Electrical and Magnetic Properties of Metals* (1976).

B-22. *Current Trends in the Field of Intellectual Property. A Series of Lectures Organized by the World Intellectual Property Organization in Montreux, June, 1971.* World Intellectual Property Organization, Geneva (1971). 399 p. Lectures on patent and trademark developments and their related data problems in OEEC, Latin America, Africa, the U.S.A., the U.S.S.R., Japan, India, and the U.K. In addition, developments under the PCT, under the generic European Patent, and new trends in plant patents and copyright laws and conventions are also covered.

B-23. Dorsey, N. E. and C. Eisenhart "On absolute measurement," *Science Monthly*, **77**, 103-109 (No. 2, 1953).

B-24. Gnaedinger, J. P. "Are we ready for standards on laboratory competence?" *Materials Research and Standards*, **12**, 8-13 (No. 11, 1972). Methods of testing as well as equipment tests need to be made to assure proper use of engineering testing laboratories.

B-25. Golashvili, T. V. "Problems of compilation and evaluation of property data of substances and materials," *J. Amer. Soc. Inform. Sci.* **24**, 57-64 (No.1, 1973). The establishment of a numerical data system intends to provide science and technology with reliable information on the properties of substances and materials and this causes a variety of new problems. The most urgent of these is the problem of developing basic methodological principles underlying the processes of data collection, evaluation, recommendation, and dissemination.

B-26. Gorozhanina, R. S., A. Y. Freedman, and A. B. Shaievitch *Standard Samples Issued in the U.S.S.R.*, U.S. National Bureau of Standards, Washington, D.C. (1971) (NBS Special Publication 260-30).

B-27. Grossnickle, L. L., ed. *An Index of State Specifications and Standards, covering those Standards and Specifications issued by State Purchasing Offices of the United States.* U.S. National Bureau of Standards, Washington, D.C. (1973). (NBS Special Publication 375).

B-28. Groot, E. H. "Unique Identifiers for Serials: An Annotated, Comprehensive Bibliography." *Serials Librarian*, **1**, 51-75 (Fall 1976). References encompass both the criteria for an ideal code system to uniquely identify serials and the development of the existing codes, with special emphasis on the CODEN and ISSN Systems.

B-29. *Guide to Standards Committees Operating Under ANSI Procedures.* American National Standards Institute, Inc., New York (1973). 40 p. Identifies the 274 committees which, in 1973, were engaged in developing and revising standards under ANSI procedures. Also lists their scopes and secretariats.

B-30. Hague, J. L., T. W. Mears, and R. E. Michaelis *Sources of Information.* U.S. National Bureau of Standards, Washington, D.C. (1965) (NBS Special Publication 260-4).

B-31. Heller, S. R. "Quality control of chemical data bases," *J. Chem. Inform. Computer Sci.* **16**, 232-233 (1976).

B-32. Hill, M. F. and J. L. Walkowicz *The World of EDP Standards.* U.S. Institute for Computer Sciences and Technology, U.S. National Bureau of Standards, Washington, D.C. (December 1976). 122 p. Two appendices. NBSIR 77-1195.

B-33. Hillebrand, W. F. "Standard methods of sampling and analysis and standard samples," *J. Ind. Eng. Chem.* **8**, 466-469 (1916).

B-34. Horowitz, E. "Transfer of measurement methodology to industry," *J. Assoc. Eng. Architects in Israel*, **35**, 21-41 (May 1976). Traces the work of the NBS materials program from the laboratory phase to the transfer and utilization of measurement technology and standards in industry.

B-35. Horwitz, W. "The establishment of official analytical methodology." *J. Chem. Inform. Computer Sci.* **17**, 97-102 (1977).

B-36. *Industrial Property, Monthly Review of the World Intellectual Property Organization.* Geneva (1961-). Covers activities and developments under the PCT as well as other national and international intellectual property agreements and conventions. Documents the work of ICIREPAT (Cooperation in Information Retrieval Among Patent Offices), especially the work of ICIREPAT's Technical Committee for Standardization (TCST).

B-37. International CODATA Conference, 4th, Tsakhcadzor, Armenian SSR, 1974. *Generation, Compilation, Evaluation, and Dissemination of Data for Science and Technology.* Pergamon Press, Oxford, New York (1975).

B-38. "International CODATA Conference on Generation, Compilation, Evaluation, and Dissemination of Data for Science and Technology, 4th, Proceedings." ***CODATA Bulletin*, 14** (1975). Covers data centers, national, and industrial, as well as CODATA's role in meeting needs of various sciences. (U. S. A. U.S.S.R., U.K., France, Japan, Poland, etc.).

B-39. International CODATA Conference, 5th, Boulder, Colo., 1976. *Generation, Compilation, Evaluation, and Dissemination of Data for Science and Technology*, Pergamon Press, Oxford, New York (1977). Includes several papers on national and international data programs. (West Germany, Japan, U.S.S.R., U.S. National Archives, UNISIST, World Federation of Engineering Organizations, Euronet, IAEA Nuclear Data Center). Also discussion on CODATA.

B-40. "International CODATA Conference on Generation, Compilation, Evaluation, and Dissemination of Data for Science and Technology, 6th, Santa Flavia, Sicily, 1978. Selected Papers." ***CODATA Bulletin*, 29** (1978).

B-41. International CODATA Conference, 6th, Santa Flavia, Sicily, 1978. *Generation, Compilation, Evaluation, and Dissemination of Data for Science and Technology.* Pergamon Press, Oxford, New York (1979).

B-42. International Council of Scientific Unions. Committee on Data for Science and Technology. *International Compendium of Numerical Data Projects: A Survey and Analysis.* Springer, New York (1969). Covers national data programs, data centers, and continuing numerical data projects.

B-43. International Federation of Library Associations (comp.) *Standardization Activities of Concern to Libraries and National Bibliographies: An Outline of Current Practices, Projects and Publications.* IFLA Committee on Cataloguing, London, (1976). 36 p. A review of the present studies of bibliographic practices by national bibliographies to enable the IFLA/UNESCO International Congress on National Bibliographies "to suggest standards for those elements which are common to all national bibliographies and to make recommendations which have validity for all." Part II gives full bibliographic citations, source, cost, and availability of all documents cited in Part I (Standards and Standard Practices in Use).

B-44. International Organization for Standardization. *Documentation-Format for Bibliographic Information Interchange on Magnetic Tape.* Geneva (1973). ISO 2709-1973. See also ISO items B-52, B-53, B-54, and B-91.

B-45. International Organization for Standardization. *Documentation-Presentation of Scientific and Technical Reports.* Geneva (1978). 31 p. Draft International Standard ISO /DIS 5966.

B-46. International Organization for Standardization. *Information and Documentation-Vocabulary Section 3: Identification, Acquisition and Processing of Documents and Data.* Geneva (1978). 23 p. Draft International Standard ISO/DIS 5127/ 111.

B-47. International Organization for Standardization. *ISO Member Bodies.* Geneva (1973). 60 p. Includes a brief description of ISO membership, responsibilities, and funding. The member standards activity or association of the 1973 56-member countries are described in one-page profiles. Text in English.

B-48. International Organization for Standardization. *Report to Council of ad hoc Group on Reference Materials.* Geneva (1974).

B-49. International Organization for Standardization, *ISO Catalogue, 1975*, Geneva (1975). 155 p. Revised and reissued annually. Preface in English, French, and Russian; text in French and English. Lists ISO publications by Technical Committee and in numeric order. Includes a separate section, listing those standards available in trilingual edition (English, French, and Russian) by Technical Committee and in numeric order.

B-50. International Standards Organization, *ISO Memento, 1975*, Geneva (1975). 176 p. Brief description of how an international standard is developed. Includes names and addresses of member bodies arranged by country, describes organization of ISO components, describes work and composition of each Technical Division and Technical Committee. Revised annually, text in English, French, and Russian. (In 1977, the ISO member bodies numbered 64).

B-51. International Standards Organization, *ISO Participation in ISO Technical Committees and Technical Divisions.* Geneva (1977). 10 p. Charts of member bodies' participation in ISO activities by country and ISO Technical Committee/Technical Division by number. Full text in English; partial text in French and Russian.

B-52 International Organization for Standardization and Unesco. *ISO Standards Handbook 1: Information Transfer*, Geneva (1977). 516 p. (ISBN 92 67 10017 3). Also available in French (ISBN 92 67 20017 9). Texts of ISO Standards. Conceived as a back-up to *UNISIST Guide to Standards for Information Handling* (B-93) (Copies available through ISO or Unesco.)

B-53 International Organization of Legal Metrology. *Concise Report of the ad hoc International Meeting on Reference Materials.* Bureau International de Metrologie Legale, Paris (1973).

B-54. International Symposium on Numerical Reference Data for Science and Technology, Warsaw, 1969. *Proceedings.* Ossolineum, Warsaw (1973). Includes information on numerical reference data programs, activities of data centers and discussions of metrological and spectral data requirements. Symposium sponsored by CODATA, the Polish Academy of Sciences and the Polish National Board for Quality Control and Measures.

B-55. International Union of Pure and Applied Chemistry. Commission on Physicochemical Measurements and Standards. "Catalogue of physicochemical standard substances," *Pure Appl. Chem.*, **29**, 599-616 (No. 4, 1972).

B-56. Kuiper, B. E. *Information Systems for the International Accessibility of Standards.* Tilburg University, Netherlands (1975). 184 p.

B-57. Landis, J. W. "ANSI's Accomplishments; Annual Report - 1977." American National Standards Institute, New York (1977). 4 p. A brief overview of the year's activities.

B-58. Lide, D. R. "The National Standard Reference Data System as a Materials Information Resource," in *Materials Information Programs.* U.S. National Bureau of Standards, Washington, D.C. (1977). (NBS Special Publication 463).

B-59. *List of Voluntary Product Standards, Commercial Standards, and Simplified Practice Recommendations* U.S. National Bureau of Standards, Washington, D.C. (1977). (NBS List of Publications No. 53).

B-60. Livingston, L. G. "Bibliographic Standards and the Evolving National Library Network." In *Academic Libraries by the Year 2000: Essays Honoring Jerrold Orne.* H. Poole, ed. Bowker, New York (1977).

B-61. Mandel, J. "Statistics and standard reference materials," In ref. B-62 (1975). p. 146-160.

B-62. Materials Research Symposium, 6th, National Bureau of Standards, 1973. *Standard Reference Materials and Meaningful Measurements.* U.S. National Bureau of Standards, Washington, D.C. (1975). (NBS Special Publication 408).

B-63. McCoubrey, A. D. "The national measurement system and specific industrial needs." In Fluid Power Testing Symposium, Milwaukee, 1975. *Proceedings.* Milwaukee, Fluid Power Society (1975). p. 1.1.1-1.1.8. An effective measurement system will meet the needs of the users in terms of acceptable and accessible central standards, reliable transfer standards, practical levels of accuracy, cost, and unambiguous language for communication.

B-64. Meinke, W. W. "Standard reference materials for clinical measurements." *Anal. Chem.*, **43**, 28A-47A (No. 6, 1971).

B-65. *Metrology and Standardization in Less-Developed Countries: the Role of a National Capability for Industrializing Economies.* U.S. National Bureau of Standards, Washington, D.C. (1971). (NBS Special Publication 359).

B-66. "The NBS Standard Reference Materials Program," *Anal. Chem.*, **38**, 627A-640A (No. 8, 1966).

B-67. *The Patent Cooperation Treaty*. Washington, D.C., U.S. Patent Office (1971). 327 p. Includes brief history and text of Treaty, plus Rules governing each Treaty Chapter, rosters of the delegates, observers and the Secretariat to the Washington Diplomatic Conference on the Patent Cooperation Treaty (May 25-June 19, 1970). Texts of Treaty and Rules in French and English. Rule 34, Chapter I, lays down minimum documentation requirements for international patent applications and international patent searches. Chapter IV of the Treaty establishes information services to facilitate acquisition of technology and technical information by developing nations.

B-68. Physikalisch-Technische Bundesanstalt. *Normalöle für Viskositätsmessungen*. Brunswick, Germany (1974).

B-69. Physikalisch-Technische Bundesanstalt (PTB). *Radioactiv Standardsubstanzen*. Brunswick, Germany (1974).

B-70. "Product standards for a summer afternoon," *Dimensions/NBS* (1975). pp. 157-159.

B-71. Radin, Nathan. "What is a standard?" *Clinical Chem.*, **13**, 55-76 (No. 1, 1967). Discusses the use of standards and the role of NBS in providing standards and reference samples.

B-72. Rather, L. J. and P. J. De La Garza "Getting it all Together: International Cataloging Cooperation and Networks." *J. Library Autom.*, **10**, 163-169 (No. 2, 1977). Traces the development of a shared international bibliographic control system from 1960-1977. Includes an account of international bibliographic standardization activity.

B-73. "Recommendations for the Presentation of Raman Spectral Data," *Appl. Spectroscopy*, **30**, 20-23 (1976). Recommendations for standards for data presentation of normal Raman scattering from isotropic materials are discussed. The types of data to be included and their proposed format are also treated.

B-74. *Reference Manual for Machine-Readable Bibliographic Descriptions* Unesco, Paris (1974). 71 p. UNISIST Publication SC. 74/WS/20.

B-75. Reventhal, A. "A national standards policy: an organizational approach." Thesis, George Washington University, School of Engineering and Applied Science (1971). Compares U.S. nongovernmental standards efforts and Canadian national policy, and discusses the alternatives.

B-76. Roberts, R. W. "The national/international measurement system," *ASTM Standardization News* **2**, 8-13 (No. 11, 1974). National and international measurements are vital to our national economy. They are vital in business transactions measuring the quality of materials and goods, vital to citizens in measuring the performance and safety of products, and vital to government for measuring compliance with national laws. This measurement system must have anchor points and its quality must be monitored and improved.

B-77. Rossini, F. D. "The ICSU Committee on Data for Science and Technology (CODATA)." *J. Chem. Doc.*, **10**, 261-264 (1970).

B-78. Rossmassler, S. A., ed. *Critical evaluation of data in the physical sciences: a status report on the National Standard Reference Data System*. U.S. National Bureau of Standards, Washington, D.C. (1977). (NBS Technical Note 947).

B-79. Shaievich, A. B., ed. *Standartnye obrazoy vypuskaemye v SSR*. Izdatelstvo Komiteta Standartov, Moscow (1969).

B-80. Slattery, W. J. *An Index of U.S. Voluntary Engineering Standards*. U.S. National Bureau of Standards, Washington, D.C. (1971). (NBS Special Publication 329). Voluntary engineering and related standards, specifications, test methods, and recommended practices published by some U.S. technical societies, professional organizations, and trade associations. Suppl. 1 (1972). Suppl. 2 (1975).

B-81. *Standardization*. United Nations, New York (1969). Based on the Proceedings of the International Symposium on Industrial Development, Athens, 1967.

B-82. Stiehler, R. D. "Standards and standardization." In *Symposium on Recent Advances and Developments in Testing Rubber and Related Materials, Philadelphia, 1973*. Philadelphia, American Society for Testing and Materials (1974). (ASTM Special Technical Publication 553). Role of engineering standards, need for unified international standards and for attention to the performance of standards testing laboratories.

B-83. Taylor, B. N. *Fundamental Physical Constants*. U.S. National Bureau of Standards, Washington, D.C. (1974). (NBS Special Publication 398). A card giving the values of the 1973 least-squares adjustment of the fundamental constants carried out by Cohen and Taylor.

B-84. Taylor, B. N. and E. R. Cohen "Present status of the fundamental constants." In International Conference on *Atomic Masses and Fundamental Constants 5*. Plenum Press, New York (1976). p. 663-673. Examines the effect of the most important relevant new experiments and calculations on the recommended output values of the 1973 least-squares adjustment.

B-85. Thompson, L. D., K. Beckman, and P. Ricci *Standards Cross-Reference List*. 2nd ed. MTS Systems Corp., Minneapolis, Minnesota (1977). Cross-reference to voluntary engineering standards adopted by two or more organizations.

B-86. United Nations Industrial Development Organization. *Standardization*. United Nations, New York (1969). 66 p.

B-87. UNISIST. *Synopsis of the Feasibility Study on a World Science Information System*. Unesco, Paris (1971). 92 p.

B-88. *U.S. Code Congressional and Administrative News, 94th Congress, 1st Session, 1975*. West Publishing Company, St. Paul, Minnesota (1976). Two vols. Volume 1 contains the texts of all public laws passed by the 1st Session of the 94th Congress (January 14, 1975-December 19, 1975). Volume 2 contains the legislative histories of the more important of the public laws included in Volume 1.

B-89. U.S. National Bureau of Standards. Office of Standard Reference Materials. *Catalog of NBS Standard Reference Materials, 1975-76 edition*. U.S. National Bureau of Standards, Washington, D.C. (1975). (NBS Special Publication 260).

B-90. UNISIST. *Guide to Standards for Information Handling* (Ex: *UNISIST Manual for Information Handling Procedures*). Unesco, Paris (in press 1979). Comprehensive guide for users and processors of information stressing evaluation of standards and specific spheres of application. Priority is given to International ISO standards (see B-52). Chapter IX treats handling of numerical data.

B-91. International Organization for Standardization and Unesco. *ISO Standards Handbook 1. Information Transfer.* Geneva (1977). 516 p. (ISBN 92 67 10017 3). Also available in French (ISBN 92 67 20017 9). Texts of ISO standards. Conceived as a back-up to *UNISIST Guide to Standards for Information Handling* (B-87). Copies available through ISO or Unesco.

B-92 King Research Inc. Center for Quantitative Sciences. *Statistical Indicators of Scientific and Technical Communications, 1960-1980, 1, A Summary Report,* U.S. Government Printing Office, Washington, D.C. (1976). 99 p. (Report prepared for the National Science Foundation.)

B-93. UNISIST. *Guide to Standards for Information Handling (Ex: UNISIST Manual for Information Handling Procedures).* Unesco, Paris (in press 1979). Comprehensive guide for users and processors of information stressing evaluation of standards and specific spheres of application. Priority is given to International ISO Standards (see B-52). Chapter 9 treats handling of numerical data.

Data Handling for Science and Technology
S.A. Rossmassler and D.G. Watson (eds.)
North-Holland Publishing Company

USE OF COMPUTERS IN HANDLING OF LABORATORY DATA

I. Eliezer

ABSTRACT

This chapter provides a very brief summary of major aspects of the computer handling of laboratory data. Topics covered include error prevention and automated editing, computerized publication, networks, interactive systems, data base management, standardization of data representation and information processing, privacy and security, copyright of machine-readable files.

OVERVIEW

The use of computers to acquire, reduce, massage, archive, access, retrieve, display and evaluate data in a variety of forms has been steadily expanding in a wide range of disciplines including physics, chemistry, geology, biology, etc. The title of this chapter has been carefully chosen since we are not attempting to consider here the use of computers in the treatment of observational data as in the astro- and geosciences. Data sets in these disciplines can be very large, requiring for storage the equivalent of 10^6 magnetic tapes. Some of the most powerful and highest capacity computers in the world are used for coping with meteorological, space science, seismological or marine survey data. The techniques are developing so rapidly that any treatment here, even if fundamental, would be out of date before this manuscript could be published. Some idea of the scope of this aspect of computer use in data handling is given in Chapter 3B.

Computer techniques for the generation of printed or otherwise typeset material have developed to a stage where such systems can deal with almost any combination of symbols and alphabets under complete computer control.

Numerous computer systems capable of accessing large centralized data banks from remote terminals via telecommunications are now operational. An alternative computerized data dissemination method is based on decentralized data banks, updated by interchange of defined format tapes.

In computations, although the batch mode process is probably more efficient overall, direct access time-shared facilities have proved invaluable where there is need to communicate with the computer during program operation or for more immediate results.

Mini-computers can handle problems of increasing complexity in data acquisition and storage, data interpolation and data base management in general.

The problems of computerized data-handling are increasingly becoming socio-economic rather than technical. The progress of such systems will, therefore, depend to a considerable extent on the way their operating costs will be shared among the generators, compilers, evaluators and users of the data.

COMPUTERIZATION OF DATA ACQUISITION, STORAGE AND RETRIEVAL

Computer system techniques for both text and numerical data systems rest on a common foundation of data organization, computer operations, and programming language.

It is convenient to divide a computer-based data-handling system for laboratory use into three major segments: (1)

- The Input Segment which includes acquisition of data sources, selection of the data sets to be entered, checking and editing the data to conform to the standards of the system, and conversion into the form necessary for storage in a data base.

- The Data Base Segment which includes the organization and storage of the data, provision of logical access to the data for storage and retrieval, provision of access rights and limitations for entry of new data, for updating of data already stored and for data retrieval, provision of processes to reconstruct file content in the event of hardware, software or human malfunction, provision of information about the data base for use in the organization, storage and retrieval processes, and provision of statistics on the use of items in the data base to aid in its management.

- The Output Segment which includes the selection of items to be output for a specific use, the organization and formatting of the data selected for delivery, and the provision of delivery means.

At a more general level, the entire process of data-handling comprises three stages: First, the generation of data from experiments and their presentation in original reports and publication.

The second stage involves the collection, condensation and organization of the data so that relevant data can be found without the necessity to go over everything that is published or compiled. This stage includes abstracting, indexing, and data center activities. It comprises the analysis and reduction of data for service to users.

The last stage involves selection and adaptation of data according to user needs and requests and its provision either routinely or on demand. This stage includes local data centers and, in fact, also information services and specialized libraries.

Each of the above stages can be computerized to varying degrees. For maximum benefits, computerization should be introduced in such a way that, once the data are entered in machine-processible form, no human action is required to carry the data through the various processes. This should eliminate costly human transcription errors. However, it requires absolute precision in those elements used to direct automatic processes at various points in the computerized system, i.e., identifiers, sortkeys, workflow controls, etc. The automatic processes will, of course, lead to erroneous results if supplied with erroneous data. For both purposes, accuracy must be attained as early in the system as possible and then maintained.

Error prevention is, of course, one of the functions of editing. Those characteristics subject to mechanical treatment are exploited for automatic editing (2).

The simplest kind of editing subject to computer checking is based on data characteristics. A particular data element may be all numeric or all alphabetic or limited to a specific range of values. In some cases the size of the data field entered may have a specific length. In a given processing step certain data elements may be required and others may be optional. However, once unique identifier codes have been assigned to data elements, their accuracy must be preserved. This is usually done by adding artificial redundancy which can be mechanically edited to check correctness, i.e., a check digit or check character appended to the basic identifier. If the data representation has natural redundancy, it can also be used for editing purposes. Within a specific set of data elements there may be rules for consistency of co-occurrence or form or specific range or values for selected subsets of elements. Applying these rules for consistency checking is another useful form of computerized editing.

In the final analysis, however, both human and machine efforts must be applied to obtain full benefit from an automated acquisition, storage and retrieval system.

COMPUTERIZATION OF DATA PUBLICATION

Present day computers are capable of processing large amounts of mathematical, physical, chemical and other data in a very short time. The equivalent of hundreds of pages of numerical or tabular data can be obtained in a few hours or even a few minutes. When the calculation or compilation or other processing is completed, the problem of publication of the data arises.

The most common output of computerized data-handling systems is still that of typing or printing. In the past several years, the interaction of computers and typesetting devices has attracted a great deal of attention, and increasingly automated techniques have emerged for use in publication and printing processes.

A technique for computerized typesetting of tables from magnetic tapes was developed very early at the NBS by Bozman (3). Several books of data have been produced by this method which entailed preparation of special programs in machine language.

Another set of pioneering programs for computer-assisted typesetting were developed at MIT by Barnett (4) producing, among other things, the famous "Tail" from "Alice in Wonderland."

More recently, the Data Systems Development Group at the NBS OSRD have developed several general programs enabling any computer user to prepare magnetic tapes for phototypesetting (5).

In general, computerized publication processes range from direct computer printouts, using high-speed but low quality printers, through computer-generated tapes that drive automatic typesetting operations to batteries of low speed composer printer devices, and computer-controlled high speed photocomposition systems providing high-quality typographic output.

Direct, on-line printouts are usually made at high speed but with monocase limitations, fixed intercharacter and interword spacing and poor utilization of paper. There are continuing trends for higher and higher speed of printer output especially in the area of non-impact printers, with rates of the order of a 1000 lines per second already attained.

The major type of computer typesetting today is computer-assisted photocomposition (6,7). Photocomposition techniques involve the transfer of optical images of characters from a negative master to a photographic emulsion, with suitable provisions for illumination, image magnification or reduction, and positioning. A photocomposing machine will accept, as input, instructions in the form of paper or magnetic tape. The images can be transferred onto film or paper from digitized information held in a computer store with a speed of about 6000 characters per second.

Microform computer output is a keen competitor to photocomposition. An advantage of microforms is the reduction of the need for remote physical access to central files because an entire file of information and, in fact, a small library can be on an individual's desk. It can be available on simple command and capable of easy update and correction of errors because of the erasability offered by thermoplastic or photosensitive microforms. However, a critical factor is the users' willingness to accept and use the products and processes of microform techniques.

With time there will probably occur a closing of the gap between photocomposition and microforms with machines concurrently producing output suitable both for printing in a size for reading without aids, and in a miniature form for people with viewers.

COMPUTERIZATION OF USER ACCESS TO DATA

The ultimate test of any data (or information) system is whether the users can quickly, easily, and comprehensively acquire the data (or information) they need. This requires, first of all, the establishment of appropriate access routes to the body of data available. Many kinds of access elements are used: citations of specific physicochemical (or other) properties, abstracts, index entries, etc. The proper access points should be use-dependent and should allow for the total intended spectrum of uses.

Once the access points to a data collection are set, the problem arises how to order and group the collection of access elements so that related items, e.g., the logical addresses in a computer file, are adjacent. To avoid the necessity to use human judgment to perform the ordering, a machine analysis can be used to derive an unambiguous ordering key, called a sortkey, so that the ordering can be done by automatic sorting.

An access route problem of specific interest to the numerical data field is augmentation of indexing of published information so that it is easier to find and retrieve numerical data and information pertaining to numerical data topics. To this end, a set of numerical data indexing principles should be derived so that the existence and type of numerical data in bodies of information is tagged explicitly and the tags included in indexes. Explicit tagging is particularly advantageous for computer searches. Such data tagging should, therefore, improve considerably user access to data. (See also Chapter 6.)

The need for ready access to diversified data banks necessitates the pooling of resources. This can be accomplished by setting up communications between independent computer data-handling systems so as to allow time-sharing, multiprocessing, and sharing of hardware, programs and data. These are the so-called computer networks (8,9). In addition to improving data access, networking also promotes greater intellectual contact and communication among users irrespective of their physical location. However, networking also presents problems of security, accuracy, organization, management, heightened requirements for good documentation, etc. Networking provides a vehicle within which solutions to current deficiencies can be worked out, not a solution itself.

The idea of a universal computerized data-handling network that would offer coherent service to users has been discussed for some time now. The problem is to predict users' future requirements, taking into account the heavy investment and the long time-scale necessary to create such a network. However, there have been various proposals for an all digital data network which might be compatible with a future digital telephone network.

A more radical approach is to consider the provision of communications facilities between people and computer systems as an overall systems problem. This has permitted the adoption of solutions based on new principles that are able to provide a data network with properties best suited to the users' requirements. Packet switching has been the result. It appeals to computer designers because its features meet their system requirements, while they can readily appreciate the principle of dynamic resource sharing on which it is based.

However, the provision of common data connection service is only one part of the solution to the problem of easy access to data and interaction between users. Just as all people have the same mechanism of speech and hearing but use different languages which prevent them from intercommunicating, the very considerable differences among computer system languages and data structures used for different computers prevent them from communicating despite the availability of universal transport mechanisms for data. The inherent incompatibility between computer systems has always been a serious problem for users, but the advent of data networks has made it much more worthwhile to find a solution.

INTERACTIVE SYSTEMS FOR COMPUTATION AND DESIGN

Not too many years ago, it was generally assumed that data users are able to assimilate only final, summarized compilations, and so computers were used primarily to process in batch and tabulate data in varied formats and content to meet the different requirements of users. However, the rapid pace of society, the energy crisis, the general concerns of environmental and socioeconomic conditions have altered the emphasis towards increased on-line interactive data management and retrieval technology so users are increasingly being offered the options of massaging data on their own and defining their own outputs as dictated by their special needs.

An interactive data-handling system makes it possible to formulate queries, carry out searches and review retrievals. The user can interact with the system by input, search and retrieval commands which are both interactive and graphic. The interactive formulation of queries must be coupled to an interactive searching capability if the system is to be utilized in real time, and if the interaction between the user and the search process is to be maximized. Because search and retrieval can be virtually instantaneous, the user's rate of hypothesis formulation can be considerably enhanced.

Scientific computation-oriented time-sharing systems often have data structure capabilities providing primarily for sequential access or access by address to regular arrays of uniform elements. Their data sharing and data recovery capabilities often place much burden on the individual user. Thus, application to interactive retrieval of large amounts of quantitative data often requires much additional systems programming. On the other hand, commercial data-oriented systems vary widely, but the largest handle up to thousands of active terminals and their storage and access capabilities allow them to deal easily with quantitative scientific and technological data. Systems which can accommodate one hundred or more simultaneously active terminals fall into two categories: those which operate through procedural language extensions, and those which are self-contained. The procedural language extensions use complex hierarchic record structures as a base representation form. Indexes for obtaining fast access to records on the basis of content are generally maintained and searched separately. In contrast, the self-contained systems generally have their own languages and provide non-procedural instructions for isolating and presenting subsets of data. These languages usually have their indexes automatically maintained and searched. However, the difficulty with these systems is that they have limited capabilities for procedural processing of the data.

Probably hundreds of types of interactive systems are presently in existence and thousands of such systems are installed. They could provide a basis for a world-wide system of storing, processing and presenting scientific and technological data. However, the economic justification of such a system is yet to be provided.

Interactive computer-user applications beyond data-handling as described above are computer-assisted instruction and computer-aided design which deserve sections of their own. Another application which is directly relevant to all the above topics as well as to the vital field of computer-assisted problem solving is the development of full-scale graphic input-output and display. At least as early as 1953 a CRT display system was available on the Iliac (10). Graphic data processing can be effectively applied in science, engineering, education, medicine, communications, business, etc. Graphic computer-aided design can be applied to projects as diverse as ship building, hospital design, flight test data reduction, cutting tool control, circuit analysis, trajectory studies and freeway interchange planning.

All the above should indicate that the future potential of computer-user interactive situations deserves vigorous R&D attention.

GENERAL-PURPOSE DATA-HANDLING COMPUTER PROGRAMS

Traditionally, computer software development involves assessment of the specific requirements which the program is to meet and design of the program so as to meet those particular requirements. Such a program has to be modified if any operational changes become necessary. Modification of the program often requires the same effort as the writing of the original program since familiarity with the program is essential. Moreover, a special-purpose program written to solve a specific problem is often no simpler than a general-purpose program which can solve not only this problem but many other related problems. Such general-purpose programs become vitally important when the need arises to cope with large numbers of diverse data files and systems. Thus, in recent years an increasing number of general-purpose programs have been developed (11-13). Such systems are called generalized data base management systems. A complete generalized data base management system can be expensive to develop and may require 30-40 man years of effort and hundreds of hours of time on a large computer for development testing (14). Nevertheless, attempts to develop such general-purpose programming systems were evident already in the mid-fifties and by the early seventies a variety were functioning in thousands of locations (15). These systems have to meet requirements for reconciliation of variable input, storage and output formats, flexible means for checking on input, combinatorial selection and output reformatting. They allow partitioning, rearranging and converting data from different compilations, restructuring and transforming files. They include self-contained query and output-formatting language for use by non-programmers and extensive updating possibilities. They can store queries for reuse and handle wide variations in the appearance and size of data elements.

General purpose data-handling programming is greatly facilitated by the use of subroutines of modular design which can be used in a variety of differing applications and can be as useful and versatile in scientific programming as in file manipulation and data retrieval. Overall, the design strategy for generalized data base management systems emphasizes the properties of both numerical and non-numerical data, and in particular the so-called "data independence" (16). To achieve this objective, data are not only stored in files according to a pattern, but descriptive information about them and their storage structure are explicitly recorded along with the data.

Although, as the above discussion shows, there is no lack of information on general-purpose data handling programs, there still remains much system design and development to do to create fully effective and economical large scale data handling software.

BIBLIOGRAPHIC AND NUMERICAL DATA BASES

Both bibliographic and numerical computerized data bases rest on a common foundation of data organization, computer system operation, and programming languages.

Bibliographic file search systems are oriented toward finding specific documents which contain desired information and toward prearranged selective capability with predetermined formatting for output. They are procedural, usually requiring updating through separate update processes and they tend to be limited to alphanumeric or string data. They usually lack the capability for complex hierarchy in inter-record relationship (17).

Unlike the considerable growth of computerized bibliographic data base systems in the past several years, computerized scientific-technological numerical data base systems have been off to a slow start. Computerized numerical data bases should not be confused with computerized bibliographic data bases whose digital numbers are indexes or references to abstracts or documents. The scientific-technological numerical data base is a digitized computer-readable record on a magnetic tape or disc so that the data can be accessed and used directly at the user's computer site or via a terminal link-up network to a remote computer. A data base system is a functioning combination of one or more data bases, and a search system which can retrieve the data - their values, and the association among them - from the data base and further process them.

Nowadays, on-line interactive bibliographic data base systems have progressed to a point where there are many such bases available on a subscription basis, e.g., the U.S. National Technical Information Service (NTIS). Techniques used to store and retrieve information for these large files (of the order of several million records) can be utilized similarly for handling large numerical data files. However, numerical data base system development has been sluggish because the problems are greater by an order of magnitude. In addition to a computerized data base, a software package is essential so as to acquire, retrieve, display, compute, manipulate and update the data. To develop good numerical data systems, one has to address input formats, data flagging and tagging, data reduction and conversion, interactive retrieval, display formats, computational and modeling capabilities, user-oriented languages and economics; costs of generating a computerized data base today are very high and cost recovery has proved uncertain.

Computerized bibliographic data bases today probably number in the thousands. These systems are nonstandardized and each has special access and procedure requirements. Thus, although it generally is agreed that, from an effectiveness viewpoint, it is highly desirable to allow the end user to engage the information system himself rather than through an intermediary, this places a heavy burden of learning on the end user. In fact, because of the above-mentioned heterogeneity of data bases and systems, placing the end user directly on-line is impractical. Two possible solutions are a) to standardize all data bases, access procedures, software and hardware of the systems as much as possible, which is a formidable task, and b) to convey the impression of standardization through the use of a computerized interface translator interposed between end users and the systems they wish to access. The interface thus creates a uniform "virtual" system and it is this single virtual system that the user engages (18).

Eventually, one can envisage the creation of integrated bibliographic and numeric data files accessible in multiple network fashion. Within a few years new technology such as value-added networks (19) should reduce data communications costs facilitating the evolution of more sophisticated combined network systems. To take advantage of those, the sophistication of the users must also increase and education and training activities should be emphasized at all levels.

STANDARDIZATION OF DATA REPRESENTATION AND INFORMATION PROCESSING

The enormous expansion in the collection, processing and exchange of data required for science and technology has made the need for standardization of data and information processing ever more urgent. Thus, considerable effort has been made in recent years to develop and apply data standards and to facilitate data processing through standard data representations. (See Chapter 8 for a discussion of related developments.)

Information processing standardization is directed to the problem of reducing existing costly incompatibilities between computer systems, networks and computer-produced information. Governmental programs are in existence in some countries to officially promulgate such standards. Thus, the U.S. Government uses the Federal Information Processing Standards Publications (FIPS PUB) Series as the publication medium for information relating to ADP standards which are adopted, as well as standardization activities nationally and internationally.

Technically, there are three methods of approaching data standardization based on the distinction among individuals, classes of individuals, and pure classes:

- the unit approach,
- the class approach, and
- the classification approach.

Common to all three approaches is the basic reliance upon the value of the unit of measuring, which is called the data variable or the data item which is listed in a code or other representational structure.

The essential problem of data standardization is that before meaningful data interchange can occur, mutual understanding must exist regarding the identification and definition of the data items involved. Such identification and definition of the codes used in the interchange, as well as description of the position or location of the data elements in the record, form the basis of data standardization. This requires that variations in the data being interchanged be eliminated or at least minimized.

The greater the agreement at the national and international levels and the more inclusive the forms of representation, i.e., the names of elements, the codes, the coding methods, and the record forms that are standardized, the more effective will be the efforts in data standardization. To this end the CODATA Task Group on Computer Use is currently studying the problem on behalf of Unesco and a survey is being conducted of existing formats. These will be analyzed and proposals will be formulated leading to an international data exchange format.

PRIVACY AND SECURITY IN COMPUTERIZED DATA SYSTEMS

In 1973 and again in 1974, the NBS sponsored two conferences on privacy and computer security, which gives some indication of the importance this problem has assumed in the U.S. and, in fact, throughout the world. What is the problem? It is based on the general recognition that data are a major asset and their availability and privacy are often critical to their proprietor. Therefore, protection must be provided from the improper disclosure, modification or destruction of the data (and necessarily also of the system), whether these are accidentally or intentionally caused. Losses from such causes have ranged into the millions of dollars. Finding ways to protect computer systems from damage, manipulation or malfunction and to protect data and the access to them is largely a scientific-technological problem. The total system must be adequately protected from unauthorized access and exploitation by others. Recovery techniques must be provided in the event of system failure or destruction of stored data.

The major approaches providing technological safeguards and procedures through which access to the systems may be controlled are based on the fundamental principle of isolation, i.e., providing mechanisms for isolating data that cannot be bypassed by the users of the system. These approaches include:

- Virtual machine systems creating an isolated environment through techniques which provide for each user a complete system tailored solely to his specification. This approach is perhaps most applicable in operations where hardware resources are shared among different organizations each with the need to protect its information from others.

- Descriptor based systems creating an isolated environment through techniques which provide the authorized user with unbounded memory space unaccessible to others. This approach is perhaps most applicable in on-line time-sharing systems. Both these approaches must be augmented with mechanisms for identifying users and authorizing their access to the system. These include methods like voiceprints, memory passwords and fingerprints. It is necessary to restrict as well as prevent access since a user may have the right to certain information but not to other information.

In cases where unauthorized access does succeed, data encryption can provide some protection. The capability to encrypt data is particularly crucial in computer networks where achieving security is a greater challenge than in stand-alone systems. By translating information through algorithms, decoding becomes difficult though not impossible. The NBS Institute for Computer Science and Technology has developed such algorithms and is making them generally available. However, better, faster encryption techniques and speedier, less costly circuits are still needed. Since encryption technology is a specialty of governments, the ultimate success of security architecture using encryption will depend on the continued willingness of the appropriate government agencies to help develop the algorithms needed to satisfy the design criteria of data processors.

COPYRIGHT FOR COMPUTERIZED INFORMATION

A major effect of technological change is that it causes ambiguities in some of the definitions of property rights that may have seemed perfectly clear before the change.

The concept of common law copyright conforms to the philosophy that each person has the right to the fruits of his creation. However, copyright protection assumes the concept of the quid pro quo of a social contract. This requires that, in return for protection of the law, the copyright holder makes a public disclosure of his work.

The issue arising with the emergence of computer-readable media was whether a copyright owner deserves compensation when his work is first computerized (encoded), or for the time it continues to be stored, or only when hardcopy is created. The specter of the replacement of technical and scientific books and journals by a computer code replicated at thousands of remote terminals held severe implications for copyright owners. In the U.S. very recently, (20) the National Commission on New Technological Uses of Copyrighted Works has prepared recommendations to be submitted to the U.S. Congress on copyright law changes with respect to computer-readable works. These recommendations are that both computer-readable data bases and computer programs in source language should be copyrighted in any tangible medium of expression. However, complete disclosure to the Copyright office of the contents of the data base or the computer program, accompanied by a usage manual, should be required. The effectiveness of discovery of infringements in the copying and unauthorized sale and use of computer-readable work needs further study.

The growth of computerized information systems makes copyright protection of the information stored in such systems desirable for two reasons: First, economic welfare is improved by reducing the need for individual contracts between each producer of information and each system operator. Second, the cost of monitoring usage to determine the proper royalty payments is low in a high-technology system which relies on computer searches.

REFERENCES

1. Gautney, G. E., Jr. and R. L. Wigington "American Chemical Society - Chemical Abstracts Service" in *"Encyclopedia of Computer Science and Technology."* Dekker, New York (1975).

2. Gehring, R. S. et al. "Automated Editing," Chemical Abstracts Service Report No. 1 (October 1972).

3. Bozman, W. R. "Phototypesetting of Computer Output," NBS Technical Note 170, Washington, D.C. (1963).

4. Barnett, M. P. *Computer Typesetting,* MIT Press, Cambridge, Massachusetts, (1965).

5. Messina, C. G. and J. Hilsenrath, "Edit-Insertion Programs for Automatic Typesetting of Computer Printout," NBS Technical Note 500, NBS, Washington, D.C. (April 1970).

6. Seybold, J. W. *"Fundamentals of Modern Composition,"* Media, Box 644, Pennsylvania 19063 (1977).

7. Lawson and Provan *"Typography for Photocomposition,"* National Composition Assoc. Arlington, Virginia (1976).

8. Roberts, L. G. and D. B. Wesler, "Computer Network Development to Achieve Resource Sharing," *Proc. AFIPS SJCC,* (1970).

9. Davies, D. W. "Principles of a Data Communications Network for Computers and Remote Peripherals," *Proc. IFIP Congr.*, Edinburgh (1968).

10. Frank, W. L. "On Line CRT Displays: User Technology and Software in On-Line Computing Systems," *Proc. Symp. Informatics,* University of California, Los Angeles (February 1965).

11. Wall, H. "GIS - Generalized Information System," *Final 1970 Census Plans*, U.S. Bureau of the Census (January 1970).

12. McClennon, R. and J. Hilsenrath, "COMBO: A General Purpose Program for Searching, Annotating, Encoding-Decoding, and Reformatting Data Files," NBS Technical Note 700, NBS, Washington, D.C. (1972).

13. Hilsenrath, J. and B. Breen "OMNIDA: An Interactive System for Data Retrieval, Statistical and Graphical Analysis, and Data-Base Management," NBS Handbook 125, NBS, Washington, D. C. (1978).

14. Meiners, E. E. "A Machine Independent Data Management System," *Datamation,* (June 1973).

15. Rosen, S. "Programming Systems and Languages," *AFIPS Proc. Spring Joint Computer Conf.*, Vol. 25, Washington, D.C. (April 1964).

16. Senko, M. E. et al., "Data Structures and Accessing in Data-Base Systems," *IBM Systems Journal* **12**, 30 (1973).

17. Olle, T. W. "A Comparison Between Generalized Data Base Management Systems and Interactive Bibliographic Systems," in *Interactive Bibliographic Search: The User Computer Interface*, D. E. Walker, ed., AFIPS Press, Montvale, New Jersey (1971).

18. Reintjes, J. F. "The Virtual System Concept of Networking Bibliographic Information Systems" in *Advancements in Retrieval Technology as Related to Information Systems*, AGARD-CPP-207 (1976).

19. Richardson, J. M. "Value Added Network Services," *Bull. Am. Soc. Inform. Sci.* **1**, 24 (1974).

20. Saltman, R. G. "Copyright in Computer Readable Works," NBS Special Publication 500-17, NBS, Washington, D.C. (October 1977).

BIBLIOGRAPHY

B-1. McClennon, R. C. and J. Hilsenrath, "COMBO: A General-Purpose Program for Searching, Annotating, Encoding-Decoding and Reformatting Data," NBS Technical Note 700, Washington, D.C. (1972).

B-2. McClennon, R. C. and J. Hilsenrath, "Reform: A General-Purpose Program for Manipulating Formatted Data Files," NBS Technical Note 444, Washington, D.C. (1968).

B-3. Thompson, R. C. and J. Hilsenrath, "SETAB: An Edit/Insert Program for Automatic Typesetting of Spectroscopic and Other Computerized Tables," NBS Technical Note 740, Washington, D.C. (1973).

B-4. Phillips, A. H. *Computer Peripherals and Typesetting*, Her Majesty's Stationery Office, London (1968).

B-5. Wollcott, N. M. and J. Hilsenrath, "A Contribution to Computer Typesetting Techniques: Table of Coordinates for Hershey's Repertory of Occidental Type Fonts and Graphic Symbols," NBS Technical Note 424, Washington, D.C. (1976).

B-6. Duncan, B. C. and D. Garvin, "Complete Clear Text Representation of Scientific Documents in Machine-Readable Form," NBS Technical Note 820, Washington, D.C. (1974).

B-7. Campey, L. H. "Generating and Printing Indexes by Computer," Aslib Occasional Publication No. **11**, London (1972).

B-8. Barrett, R. and B. J. Farbrother, "FAX-A Study of Principles, Practice and Prospects for Facsimile Transmission in the UK," British Library Research and Development Report No. 5257HC (1976).

B-9. Terry, B. "The Systems Approach to Computer Output Microfilm," AGARD-LS-85, 51 (1976). *

B-10. Avedon, D. M. "Micrographies and COM, A State-of-the-Art and Market Report," AGARD-LS-85, 1.1 (1976).*

B-11. Spigai, F. G. and B. B. Butler, "Micrographics," *Ann. Rev. Inform. Sci. Tech.* **11**, 59 (1976).

B-12. Lykos, P., ed. "Computer Networking and Chemistry." American Chemical Society Symposium Series 19, Washington, D.C. (1973).

B-13. Pratt, G. and S. Harvey, eds. "The On-Line Age--Plans and Needs for On-Line Information Retrieval." *Proceedings of the EUSIDIC Conference, Oslo, 1975* (1976).

---

*AGARD Publications may be purchased in Microfiche or Photocopy form from National Technical Information Service (NTIS), 5285 Port Royal Road, Springfield, Virginia 22151, U.S.A.

B-14. Claydon, C. R. "Computer Software for Retrieval and Analysis of Numeric Data," *Proceedings of the 38th Annual Meeting of the Amer. Soc. Inform. Sci.* (1975).

B-15. Fried, J. B., J. A. Luedke Jr., and S. A. Rubin, "Online Numeric Data Base Systems," *Online* **1** (1977).

B-16. Vernimb, C. O. "The European Network for Scientific, Technical, Economic and Social Information," *Nachr. Dokument.*, **28**, 11 (1977).

B-17. Marron, B. and D. Fife, "Online Systems," in *Ann. Rev. Inform. Sci. Tech.* **11**, M. E. Williams, ed., American Society for Information Science, Washington, D.C. (1976).

B-18. "Feature Analysis of Generalised Data Base Management Systems," CODASYL Systems Committee (1971).

B-19. Dominick, W. D. "Numeric Data Base Systems: Concepts and Capabilities," *Collection of Papers presented at the 6th Mid-Year Meeting of the American Society for Information Science* (1977).

B-20. Davis, M. S. "Standards, Management and Security of Astronomical Data Sets," *Proc. IAU Colloq.* **35**, 1976 (1977).

B-21. Fong, E., J. Collica, and B. Marron, "Six Data Base Management Systems: Feature Analysis and User Experiences," NBS Technical Note 887, Washington, D.C. (1975).

B-22. Williams, M. E., ed., "Cost Elements and Charge Bases in Information Centers," *Proceedings of a Panel Discussion, ASIDIC Meeting March 1973* Cooperative Data Management Committee, Athens, Georgia (1974).

B-23. Woods, B. L. "Review of Scientific and Technical Numeric Data Bases," King Research, Inc., Rockville, Maryland (1977).

B-24. Bloch, M., J. Budil, I. Ganicky, and M. Svoboda, "Socialist Countries: Communicative Format of Data Recording on Magnetic Tape," *J. Chem. Info. Comp. Sci.* **17**, 32 (1977).

B-25. Benkovitz, C. M., B. N. McNeely, and R. A. Wiley, "User's Guide for the IWGDE Level 1 Implementation of the Proposed American National Standard Specifications for an Information Interchange Data Descriptive File," Los Alamos, New Mexico (1977).

B-26. "Computer Security," *Aslib Bibliography Series*, London (1976).

B-27. Hoffman, L. J., ed., *"Security and Privacy in Computer Systems,"* Melville Publ., Los Angeles, California (1973).

B-28. Feistel, H. "Cryptography and Computer Privacy," *Sci. Amer.*, **228**, 15 (1973).

B-29. Linden, B. L. "Copyright, Photocopying and Computer Usage," *Bull. Amer. Soc. Inform. Sci.* **1**, 12 (1975).

B-30. See Reference No. 13.

B-31. Luedke, J. A., Jr., G. J. Kovacs, and J. B. Fried, "Numeric Data Bases and Systems." in *Ann. Rev. Inform. Sci. Tech.* Vol. **12**, M. E. Williams, ed., Knowledge Industries Publications, White Plains, New York (1977).

Data Handling for Science and Technology
S.A. Rossmassler and D.G. Watson (eds.)
North-Holland Publishing Company

ACCESSIBILITY AND DISSEMINATION OF DATA

D. G. Watson

ABSTRACT

This chapter outlines some of the mechanisms used to access and disseminate data and attempts to indicate some of their limitations. Topics discussed include the "invisible college," synoptic journals, data depositories, handbooks, data banks, referral centers and the abstracting and indexing of data.

The dissemination of scientific information, including data, by direct contact with professional colleagues is, presumably, the oldest communication channel in science. Whereas such contact was formerly effected by letter writing it is now principally achieved by the opportunity to meet colleagues at conferences and symposia.

This development has been, in some cases, formalized to the extent that learned societies have formed special-interest groups. These groups do not necessarily have an official journal of their own but each group does provide a focus for specialist discussions, often on the occasion of larger, more general conferences.

The formal or informal network of subject specialists has, for some time, been referred to as the "invisible college." One of its activities, which has been both endorsed and criticised, is the circulation of preprints among members of the college. Provided that the subject scope is fairly limited this mechanism of data dissemination can be reasonably successful but as subject boundaries became wider and more diffuse then the general utility of the system may be questionable.

While direct contact between scientists affords an important communication channel, especially for current activities, the principal modes of accessing and disseminating data are via the published literature and information centers. The published literature encompasses a great variety of sources including journals, textbooks, handbooks, data sheets, data compilations, encyclopaedias, dictionaries, trade literature, reports, standards, specifications, abstract journals, indexes, etc. It is obviously impractical to discuss all of these in great detail but this chapter will attempt to identify some of the problems associated with both the published literature and information centers. The reader should also consult Chapters 5 and 6 for other aspects of the published literature.

The use of the primary journal literature continues to be a major method of accessing data. An obvious advantage of this literature is that it is publicly available and since, for the majority of journals, all papers have been subjected to the scrutiny of referees, a reasonable measure of reliance can be placed upon data published therein. The fact that a journal article should ideally present sufficient details of the derivation of the data to allow for a critical assessment is a decided advantage. On the other hand, in multi-disciplinary projects a particular individual may feel unqualified to make a true assessment and for this type of user the compilations of evaluated data offer a more satisfactory solution.

A major problem facing the publishers of journals is the economic one whereby increasing costs of paper and labour coupled with the growing practice of photocopying is reducing the numbers of subscribers and thus forcing up the unit cost for each journal. In the United States the basic concept of requiring an author to present his key results in one or two pages in a directly usable form was put into practice in 1971 by the American Institute of Aeronautics and Astronautics (AIAA) (1,2). This new, concise form has been called the Synoptic. The backup full papers are not published but are available to requestors at moderate cost. In another venture of this sort, the Chemical Society (London), with Gesellschaft Deutscher Chemiker and Société Chimique de France as joint and equal owners, began publication of *The Journal of Chemical Research* in a synopsis/microform concept in 1977 (3-4). Typeset synopses are published in *J. Chem. Research (S)* and the full texts (typewritten) are published concurrently in *J. Chem. Research (M)*, which is available in both microfiche and miniprint forms. Other examples of the use of synoptic journals include *Engineering Synopses* (5), *Chemie-Ingenieur-Technik* (3), *Studia Biophysica, Zhurnal Fizicheskoi Khimii* (6). Additionally the American Chemical Society has evaluated synoptic and two different full text versions of *The Journal of the American Chemical Society* (7).

Many people have complained of the added bookkeeping involved in retrieving deposited data but it is surely preferable that data should be archived under the aegis of responsible organizations rather than remain "buried" in the author's notebooks. The argument for using such deposition schemes would also appear to be very reasonable in the case of essentially raw experimental data which are often only of relevance to those who wish to repeat the scientific experiment.

An interesting and very valuable addition to the roster of journals is the data journal, especially where it is concerned with the publication of evaluated data. Such a journal was launched in 1972 under the title of *Journal of Physical and Chemical Reference Data.* It is a cooperative venture of the National Bureau of Standards, the American Chemical Society and the American Institute of Physics and its primary objective is to provide a medium for the publication of evaluated data compilations where the systematic data have been drawn from the general scientific literature.

User studies (8,9) indicate that very many people use handbooks in their search for data. For the physical sciences the one most commonly cited is the *Handbook of Chemistry and Physics*, one of a series of handbooks published by the Chemical Rubber Company of Cleveland, Ohio, U.S.A. The Biological Handbooks Series, published by the Federation of American Societies for Experimental Biology, covers a wide variety of subjects in the life sciences. A frequent criticism of some handbooks is that the data values are not updated with sufficient regularity and, in some cases, there may be insufficient indication of the reliability of the data. A few simple guidelines for the preparation of handbooks should include the following: the sources of data should be clearly stated; an indication should be available whether tabulated data are complete reproductions or selected portions of the original; standard errors should be noted, wherever possible; footnotes should not involve jargon and a multilingual listing of terms should be provided.

The major abstracts journals such as *Chemical Abstracts, Biological Abstracts, Bulletin Signalétique*, etc. have, for many years, provided indispensable tools for searching the world's scientific and technical literature. Nowadays many of these journals are printed as output from a computerised database and are also available to the user in machine-readable form. Apart from economic and time-saving benefits computerisation of these operations has enabled the abstracting/indexing services to offer specialized products to various user groups. Another important development which has perhaps been catalysed by computerisation is the attention being given to more systematic and carefully controlled indexing.

In recent years various groups have studied the problems associated with data flagging/tagging, i.e., the provision in an abstract of some indication as to what types of data are contained in the original article. This topic is discussed in Chapter 6.

The use of information centers to access data is both a long-standing and also a very new method if one agrees that the term "information center" covers both a library and an information analysis center. It is, perhaps, useful at this point to review the wide range of information-related activities which are encompassed by the term "information center." In examining the broad concept it is possible to observe that the services provided by different kinds of information centers serve entirely different purposes.

There is one family of centers which focuses on archival records; that is, they provide information on who has done what (in the past). This family consists of libraries and document centers. Some of the latter are concerned exclusively with technical reports. Both libraries and other document centers maintain extensive indexes based on broad subject categories and on authors' names. Other types of indexes may also be used. This family of centers typically replies to a request by providing one or more complete documents (books, journals, or reports) which treat some aspect of the subject indicated by the requestor.

The second family of centers provides a somewhat more personal response. They focus on who knows what. Included here are referral centers and current research activity registry centers. In response to a request, they will identify a person or organization registered as having specialized expertise in, or actively pursuing, some particular kind of knowledge.

The third category of information centers is the one most relevant to the subject of this ***Sourcebook***. It includes those centers which focus on a selected body of numerical information, and apply expert knowledge to that information in order to increase its utility and relevance. This family consists of data evaluation (or information analysis) centers, and data manipulation (or computer and calculation) centers. These centers seek to tell the requestor what the facts are.

All of these centers are useful, and in truth they are mutually interdependent. They may, indeed, be taken as forming a spectrum of activities and supplying a spectrum of answers.

• Library	Who has done what?
• Document Center	What has been done?
• Referral Center	Who knows what?
• Current Research Registry Center	What is being done?
• Data Evaluation Center	What are the facts?
• Data Manipulation Center	What is their reliability?

Over the past 10-20 years the dissemination of data has been further enhanced by the advent of data banks or numerical databases. These computerised data collections offer certain attractive features but, at the same time, they are associated with a variety of new problems for the user.

Factors which have favoured the rapid growth in the number of data banks would include the following: the sheer volume of data which is published each year necessitates that, in certain subject-areas, the only practical way of collating the data is to hold the material in machine-readable form; if the data are to be evaluated, then this very often involves extensive calculations requiring computers; as computer technology has advanced so the costs of minicomputers and on-line storage have decreased; mission-oriented projects very often require the searching of data collections from a wide range of subject-areas and this becomes a feasible proposition if the data sources are computerised.

Conversely, there are a number of factors which create problems in using data held in a computerised data bank. These include: in some cases the back-up documentation does not provide a clear indication of what kind (if any) of evaluation process has been applied to the data; in some cases the data bank does not contain the error estimates which should be associated with the data and, in the absence of the original publication, this can be a severe drawback; the searching of the data bank often involves learning a complex set of instructions, sometimes in a foreign language; the economics of searching can be a hindrance in that numeric databases often have a more limited user population than bibliographic databases and hence the unit search cost tends to be higher; this latter problem is compounded by the fact that many organisations, especially academic, are reluctant to assign real money for the computer acquisition of data and other information.

The bewildering number of data sources which are now available makes the need for efficient referral centers more urgent.

The functions of a data referral center include the following:

- to collect information on data resources, relating to data generation, evaluation and compilation
- to prepare a comprehensive file on the kinds of data available from these resources with a detailed subject index for data access
- to guide users to the appropriate resources where they may find the required data.

Many libraries and other information centers perform referral activities but there are relatively few referral centers as such. A few examples are the National Referral Center in Washington, D.C. (10), SOS-DOC in Paris, and INFOTERRA, formerly the UNEP International Referral System in Nairobi.

Recently the CODATA/ADD Task Group conducted a feasibility study (11) on the need for a world data referral center and, with the initial help of Unesco, a center has now been established in Paris (in the same building as CODATA). This World Data Referral Centre (WDRC) aims to provide a strong back-up to existing local referral services. In planning the organisation of this center due attention is being paid to a set of guidelines recently prepared by Unesco (12). One of the principal aids to the successful operation of such a center is the acquisition and regular updating of directories of information resources. Already several organizations have prepared directories to bibliographic databases, e.g. ASIDIC, EUSIDIC, Aslib and presently the EEC. Another important tool will be the updated version of the CODATA Compendium (13). Work is in hand to update this compilation on a subject-by-subject basis and the first chapter has been published (14).

The total array of data, whether in conventional printed form or in machine-readable form, is so vast that unless it is adequately indexed then it might as well be lost.

A given piece of numerical data generally refers to the magnitude of some quantity characterizing some property or phenomenon of a certain system, measured under a certain condition. The classification and indexing of data may be regarded as an elaboration of these facets. In certain areas the faceted indexing of data should be feasible, e.g., property data of pure chemical substances. However in many cases the description of the system is very complex, e.g., chemical mixtures, degree of imperfection in crystals, biological and geological locations and origins.

The definition of chemical substances, while capable of systematisation, is complicated by the problems of nomenclature. The Chemical Abstracts Service (CAS) and IUPAC have developed nomenclature rules (15-17) which provide an authoritative basis for nomenclature. Additionally various efforts have been made to develop chemical notation systems (18) aimed at handling chemical descriptions in computer form. Of these the system which has gained most wide-spread use is the Wiswesser Line Notation (19). The alternative technique which is most likely to gain general acceptance is the connectivity approach (18,21) whereby a chemical structure is described in graphic terms by the nodes (atoms) and links (bonds). This technique has been pursued by CAS as part of its Registry System (22). Each chemical substance is represented by a unique chemical graph and to each substance is assigned a unique number - the registry number. To date nearly 4 000 000 compounds have been registered by CAS. If each substance-oriented data bank contained the registry numbers of all compounds in the bank then it would be feasible to readily collate the different properties of chemical substances.

The indexing of properties of substances is an area requiring considerable expertise and experience. CAS has estimated that some 1100 terms are required to cover the range of chemical and physical properties. This statistic serves to confirm the problems faced by the Abstracting and Indexing Services in attempting to devise flagging and tagging schemes (see Chapter 6).

The multitude of problems associated with the handling of data and other information are concisely summarised in a stimulating article by Beer (20) where he states:

> *"...Thus I agree that the problem of information management is now a problem of filtering and refining a massive overload...We might well say that it is a problem not so much of data acquisition as of right storage; not so much of storage as of fast retrieval; not so much of retrieval as of proper selection; not so much of selection as of identifying wants; not so much of knowing wants as of recognizing needs - and the needs are precisely the requirements of systematic equilibrium."*

REFERENCES

1. Dugger, G. L., R. F. Bryans, and W. T. Morris, Jr., "AIAA Experiments and Results on SDD, Synoptics, Miniprints and Related Topics." *IEEE Trans. on Professional Communication, PC-16*, **100**, 178 (1973).

2. Dugger, G. L. "What Journal Changes Would Best Serve You While Countering Rising Publication Costs?" *Astronautics and Aeronautics*, **10**, 56 (1972).

3. O'Sullivan, D. A. "Synopsis Journal Idea Catching on in Europe." *Chem. Eng. News* **53**, 14 (1975).

4. *J. Chem. Research* "Instructions for Authors," Editorial Office, The Chemical Society, Burlington House, Picadilly, London W1V OBN, U.K. (1976).

5. Lambie, J. H. *Engineering Synopses, an Experiment in Synoptic Publishing.* University of Bath, Claverton Down, Bath BA2 7AY, U.K. (1976).

6. Jacolev, L. Letter to the Editor, *Chem. Eng. News* **53**, 3 (1975).

7. Tarrant, S. W. and L. R. Garson "Evaluation of a Dual Journal Concept." American Chemical Society, Washington, D.C. (1977).

8. Weisman, H. M. "Needs of American Chemical Society Members for Property Data." *J. Chem. Doc.* **7**, 9 (1967).

9. Slater, M., A. Osborn, and A. Presanis "Data and the Chemist." Aslib Occasional Publication No. **10** London (1972).

10. McFarland, M. W. "The National Referral Center." *Special Libraries* **66**, 126 (1975).

11. "Feasibility Study of a World Data Referral Centre." ***CODATA Special Report*** No. **2** (1975).

12. "UNISIST Guidelines for Establishing and Developing Referral Centres for Users of Information." Prepared with the collaboration of Mr. A. Dulong. Unesco, Paris (1978).

13. ***International Compendium of Numerical Data Projects.*** Springer-Verlag, Berlin (1969).

14. "CODATA Directory of Data Sources for Science and Technology. Chapter 1. Crystallography." ***CODATA Bulletin*** No. **24** (1977).

15. "Naming and Indexing of Chemical Substances for Chemical Abstracts during the Ninth Collective Period (1972-1976)." Amer. Chem. Soc., Columbus, Ohio (1973).

16. "IUPAC Nomenclature of Organic Chemistry. Sections A, B, C." Butterworths, London (1969).

17. "IUPAC Nomenclature of Inorganic Chemistry, 2nd Edition. Definitive Rules 1970." Butterworths, London (1971).

18. Rush, J. E. "Status of Notation and Topological Systems and Potential Future Trends." *J. Chem. Info. Comp. Sci.* **16**, 202 (1976).

19. Smith, E. G. and P. A. Baker "The Wiswesser Line-Formula Chemical Notation (WLN)." 3rd Edition. Chemical Information Management, Cherry Hill, New Jersey (1976).

20. Beer, S. "Managing Modern Complexity. The Management of Information and Knowledge." Committee on Science and Astronautics, U.S. House of Representatives (1970).

21. Dubois, J. E. and A. Panaye "Système DARC. XXI -- Théorie de génération description. X. Représentation des boranes et leur dérivés." *Bull. Soc. Chem. Fr.* 1229 (1976).

22. Dittmar, P. G., R. E. Stobaugh, and C. E. Watson "The Chemical Abstracts Service Chemical Registry System. I. General Design." *J. Chem. Info. Comp. Sci.* **16**, 111 (1976).

BIBLIOGRAPHY

B-1. Luberoff, B. J. "The Journal and its Possible Future" *J. Chem. Info. Comp. Sci.* **16**, 193 (1976).

B-2. Holm, B. E., M. G. Howell, H. E. Kennedy, J. H. Kuney, and J. E. Rush "The Status of Chemical Information." *J. Chem. Doc.* **13**, 171 (1973).

B-3. Labin, E. "Etude sur la Typologie des Banques de Données." Edition de la Documentation Française (1977).

B-4. Speight, F. Y. "Numerical Data Activities of Engineering Societies." *J. Chem. Doc.* **7**, 26 (1967).

B-5. "Geological Data Files: Survey of International Activity." ***CODATA Bulletin*** No. **8** (1972).

B-6. Wilkins, G. A. "Numerical Data Problems in Astronomy." ***CODATA Bulletin*** No. **15** (1975).

B-7. Appleyard, R. K. "The EURONET Project of the European Communities." *The Information Scientist* **10**, 159 (1976).

B-8. "Guidelines on the Conduct of a National Inventory of Scientific and Technological Information and Documentation Facilities." Unesco-UNISIST. SC/75/WS/28. Unesco, Paris (1978).

B-9. "Science and Technical Information. Why? Which? Where?and How?" AGARD-LS-44-71 (1971).*

B-10. "How to Obtain Information in Different Fields of Science and Technology. A User's Guide." AGARD-LS-69 (1974).*

B-11. Desvals, H. *Comment Organiser ca Documentation Scientifique* 2nd Edition. Gauthier-Villars, Paris (1978).

B-12. "Broad System of Ordering (BSO)." Unesco, Paris (1978).

B-13. Conrad, C. C. "Status of Indexing and Classification Systems and Potential Future Trends." *J. Chem. Info. Comp. Sci.* **16**, 197 (1976).

B-14. Wipke, W. T., S. R. Heller, R. J. Feldmann, and E. Hyde, eds., *Computer Representation and Manipulation of Chemical Information.* Wiley-Interscience, New York (1974).

B-15. Lynch, M. F. *Computer-Based Information Services in Science and Technology - Principles and Techniques.* Peter Peregrinus Ltd., Stevenage, U.K. (1974).

---

*AGARD Publications may be purchased in microfiche or photocopy form from National Technical Information Service (NTIS), 5285 Port Royal Road, Springfield, Virginia 22151, U.S.A.

B-16. Laurent, J. "La banque automatique de données terminologiques de l'AFNOR: NORMATERM" *Courier de la Normalisation* (Fr.) No. **245**, 482-486 (1975).

B-17. COMET, "Banque d'information métrologique" *Bull. Info. Bureau National de Métrologie* (Fr.) (1975).

B-18. Devoge, J. "Systeme ARIANE d'informations sur le bâtiment" *Information et Documentation* (Fr.) No. **2**, 25 (1975).

B-19. Labin, E. *Les banques de données dans le domaine scientifique et technique*, Documentation Française, Paris (1976). 338 pp.

B-20. THERMODATA. "Banque de données thermodynamiques chimiques" THERMODATA, St. Martin d'Hères, France.

B-21. CODATA-BNIST. "Utilisation de données. Banques de données dans le domaine scientifique, technique, technico - économique." Journées de reflexion, Paris (September 1977).

Data Handling for Science and Technology
S.A. Rossmassler and D.G. Watson (eds.)
North-Holland Publishing Company

APPENDIX I

NATIONAL AND INTERNATIONAL DATA PROGRAMS

This Appendix provides identification of national and international programs which deal specifically with scientific and technical data. The "title" of each is given, together with an address to which inquiries can be sent and, where available, citation of a published description of the program. Included are national programs with broad coverage (especially coverage which corresponds to the range of CODATA activities and membership), internationally-structured data projects on specific topics associated with CODATA or one of the CODATA-affiliated scientific unions, and comprehensive international organizations with a strong data orientation. In several cases, the "national program" identified is the National Committee for CODATA, which in fact provides the major national focus for a number of individual data projects.

National Data Programs

1. Australian National Committee for CODATA

   Address: Dr. G. K. White
   CSIRO National Measurement Laboratory
   P.O. Box 218
   Lindfield NSW 2070, AUSTRALIA

2. Canada Institute for Scientific and Technical Information

   Address: National Research Council of Canada
   Montreal Road, Building M-55
   Ottawa, Ontario K1A 0S2, CANADA

   Reference: ***CODATA Newsletter*** No. **14** (June 1975)
   CODATA Secretariat, Paris

3. Bureau National de l'Information Scientifique et Technique (BNIST)

   Address: Dr. Jacques Michel
   8 rue Crillon
   75194 Paris CEDEX 4, FRANCE

   Reference: *BNIST Annual Report 1973*
   (Also succeeding years)

4. National Committee for CODATA - German Democratic Republic

   Address: Professor W. Schirmer, Chairman
   Zentralinstitut für Physikalische Chemie
   Rudower Chaussee 5
   1199 Berlin-Adlershof
   GERMAN DEMOCRATIC REPUBLIC

Reference: *Proceedings of the Fourth International CODATA Conference*
Pergamon Press, Oxford (1975) or
CODATA Secretariat, Paris

5. Information System for Physics Data

Address: Dr. H. Behrens or Dr. G. Ebel
Fachinformationszentrum Energie, Physik, Mathematik GmbH
7514 Eggenstein-Leopoldshafen-2
FEDERAL REPUBLIC OF GERMANY

Reference: *Proceedings of the Sixth CODATA Conference,*
Pergamon Press, Oxford (1979)

6. Indian National Committee for CODATA

Address: Professor C. N. R. Rao, Chairman
Solid State & Structural Chemistry Unit
Indian Institute of Science
Malleswaram, Bangalore 560012, INDIA

Reference: ***CODATA Newsletter*** No. **14** (June 1975)
CODATA Secretariat, Paris

7. Israel National Committee for CODATA

Address: Dr. A. S. Kertes
Institute of Chemistry
The Hebrew University
Jerusalem, ISRAEL

Reference: ***CODATA Newsletter*** No. **19** (September 1978)
CODATA Secretariat, Paris

8. Italian National Committee for CODATA

Address: Professor Marcello Carapezza, Chairman
University of Palermo
Istituto de Geochimica
Via Archirafi 36
90123 Palermo, SICILY

9. Japan National Committee for CODATA

Address: Professor Takehiko Shimanouchi, Chairman
College of Information Sciences
The University of Tsukuba
Sakura, Niihari, Ibaraki 300-31, JAPAN

Reference: *Proceedings of the Fifth Biennial CODATA Conference,*
Pergamon Press, Oxford (1977)

10. Netherlands National Committee for CODATA

Address: Dr. W. M. Smit, Chairman
Fysisca-Chemisch Institut
TNO
Zeist, NETHERLANDS

11. Polish National Committee for CODATA

Address: Dr. Andrej Bylicki
Institute for Physical Chemistry
ul. Kasprazaka 44/52
Warsaw, POLAND

12. Swedish National Committee for CODATA

Address: Dr. Björn Tell
Director of Libraries
University Library of Lund
Box 1010 S-221 03
Lund, SWEDEN

13. USSR State Standard Reference Data System 1

Address: Professor V. V. Sytchev
Academy of Sciences of the U.S.S.R.
Soviet National Committee on Data for Science and Technology
14 Leninsky Prospect
Moscow B-17, U.S.S.R.

Reference: *Proceedings of the Fifth Biennial CODATA Conference*, Pergamon Press, Oxford (1977)

14. United Kingdom Data Program

a) for scientific data:

Address: Data Compilation Committee
Science Research Council
State House, High Holborn
London WC1R 4TA, UNITED KINGDOM

b) for industrial data:

Address: Dr. J. Sutton
Department of Industry
Abell House
John Islip Street
London SW1P 4LN, UNITED KINGDOM

15. U.S. National Committee for CODATA

Address: U.S. National Committee for CODATA
National Academy of Sciences
2101 Constitution Avenue, N.W.
Washington, D.C. 20418, U.S.A.

16. U.S. National Standard Reference Data System

Address: Dr. David R. Lide, Jr.
Office of Standard Reference Data
National Bureau of Standards
Washington, D.C. 20234, U.S.A.

Reference: NBS Technical Note 947, *Status Report of the National Standard Reference Data System - 1977* (available from above address)

or *Proceedings of the Fourth International CODATA Conference*, Pergamon Press, Oxford (1975) or
CODATA Secretariat, Paris.

Internationally-Sponsored Data Projects on Specific Topics

1. International Association for the Properties of Steam

Contact: Dr. Howard J. White, Jr.
c/o Office of Standard Reference Data
National Bureau of Standards
Washington, D.C. 20234, U.S.A.

2. International Union of Crystallography - Structure Reports

Contact: Professor J. Trotter
Department of Chemistry
University of British Columbia
2075 Westbrook Place
Vancouver V6T 1WS, CANADA

3. International Union of Crystallography - International Tables for X-Ray Crystallography

Contact: Professor T. Hahn
Institut für Kristallographie der Technische Hochschule
Templergraben 55
Aachen, FEDERAL REPUBLIC OF GERMANY

4. International Union of Pure and Applied Chemistry - Solubility Data Project

Contact: Dr. A. S. Kertes
Institute of Chemistry
The Hebrew University
Jerusalem, ISRAEL

Reference: *Proceedings of the Fifth Biennial International CODATA Conference*, Pergamon Press, Oxford (1977)

5. International Union of Pure and Applied Chemistry - Critical Surveys of Stability Constants of Metal Complexes

Contact: IUPAC Analytical Chemistry Division Commission
V.6: Equilibrium Properties

Chairman: Professor George H. Nancollas
Department of Chemistry
State University of New York, Buffalo
Buffalo, New York 14222, U.S.A.

Reference: *IUPAC Information Bulletin* No. 49, March 1975,
IUPAC Secretariat, Oxford

6. International Union of Pure and Applied Chemistry - Thermodynamic Tables Project

Contact: Dr. S. Angus
IUPAC Thermodynamic Tables Project Centre
Department of Chemical Engineering and Chemical Technology
Imperial College of Science and Technology
London SW7 2BY, U.K.

Reference: ***CODATA Newsletter*** No. **12** (March 1974)
CODATA Secretariat, Paris

International Programs of Broad Scope

1. Committee on Data for Science and Technology (CODATA)

Address: Mme. P. Glaeser
CODATA Secretariat
51 Boulevard de Montmorency
75016 Paris, FRANCE

Reference: See Chapter 1.

2. International Atomic Energy Agency (IAEA) Nuclear Data Center

Address: Dr. Alex Lorenz
Nuclear Data Section
International Atomic Energy Agency
Kärntner Ring 11
A-1010 Vienna, AUSTRIA

Reference: *Proceedings of the Fifth Biennial CODATA Conference*, Pergamon Press, Oxford (1977).

3. United Nations Environmental Program (UNEP)

   a. International Referral System for Sources of Environmental Information (IRS)

   Address: Mr. A. Koshla
   Director, INFOTERRA (International Referral System)
   P. O. Box 30552
   Nairobi, KENYA

   b. Global Environmental Monitoring System (GEMS)

   Address: Mr. F. Sella
   Director, GEMS
   United Nations Environmental Programme
   P.O. Box 30552
   Nairobi, KENYA

   c. International Register of Potentially Toxic Chemicals (IRPTC)

   Address: IRPTC Unit
   World Health Organization Headquarters
   1211 Geneva 27, SWITZERLAND

4. Marine Environmental Data and Information Referral System (MEDI)

   Address: Mr. A. Tolkachev
   Mr. M. Pobukovsky
   IOC Secretariat, Unesco
   7 Place de Fontenoy
   75700 Paris, FRANCE

5. UNISIST (Programme for Cooperation in the Field of Scientific and Technical Information)

   Address: Division of the General
   Information Programme
   Unesco
   7 Place de Fontenoy
   75700 Paris, FRANCE

   Reference: See Appendix II.

6. World Data Center System (Solar and Geophysical Data)

   Address: a) World Data Center A
   Coordination Office
   National Academy of Sciences
   2101 Constitution Avenue
   Washington, D.C. 20418, U.S.A.

   b) World Data Center B
   Molodezhnaya 3
   Moscow 117 296, U.S.S.R.

c) For address of World Data Center C, write for Consolidated Guide listed below.

Reference: *Proceedings of the Fourth International CODATA Conference*, Pergamon Press, Oxford (1975) or CODATA Secretariat, Paris.

also *Third Consolidated Guide to International Data Exchange through The World Data Centers*, issued by the Secretariat of the ICSU Panel on World Data Centers, National Academy of Sciences, Washington, D.C. (1973).

7. Federation of Astronomical and Geophysical Services (FAGS)

Address: Dr. G. A. Wilkins
Royal Greenwich Observatory
Herstmonceux Castle, Hailsham
East Sussex BN27 1RP, ENGLAND

Data Handling For Science and Technology
S.A. Rossmassler and D.G. Watson (eds.)
North–Holland Publishing Company

APPENDIX II

THE UNESCO GENERAL INFORMATION PROGRAMME

UNISIST PROGRAMME ACTIVITIES

Information transfer has long been seen as an essential element in the rational utilization of natural and human resources within the framework of Unesco (United Nations Educational, Scientific and Cultural Organization). Indeed, Article I of the Unesco Constitution stipulates that the Organization shall

> *"maintain, increase, and diffuse knowledge... by encouraging cooperation among the nations in all branches of intellectual activity... the exchange of publications... and other materials of information; and by initiating methods of international cooperation calculated to give the people of all countries access to the printed and published materials produced by any of them."*

In view of the particularly vital importance of scientific and technological information in the development of nations, a long-term intergovernmental programme entitled UNISIST was initiated by Unesco in 1973 (Resolution 2.131, 17th Unesco General Conference, 1972), following a five-year joint study with the International Council of Scientific Unions (ICSU) (1). The ultimate goal of UNISIST is a loosely connected world network of existing and future information systems, based on their voluntary cooperation; certain broad principles originally set out in the Unesco/ICSU study report are helpful in understanding the aims of UNISIST:

- the unimpeded exchange of published or publishable scientific information and data among scientists of all countries;
- hospitality to the diversity of disciplines and fields of science and technology as well as to the diversity of languages used for the international exchange of scientific information;
- promotion of the interchange of published or publishable information and data among the systems, whether manual or machine, which process and provide information for the use of scientists;
- the cooperative development and maintenance of technical standards in order to facilitate the interchange of scientific information and data among systems;
- promotion of compatibility between and among information processing systems developed in different countries and in different areas of the sciences;
- promotion of cooperative agreements between and among systems in different countries and in different areas of the sciences for the purpose of sharing workloads and of providing needed services and products;
- assistance to countries, both developing and developed, wishing access to contemporary and future information services in the sciences;

- the development of human and information resources in all countries as necessary foundations for the utilization of machine systems;
- the increased participation of scientists in the development and use of information systems, with particular attention to the involvement of scientists in the evaluation, compaction, and synthesis of scientific information and data;
- the involvement of the coming generation of scientists in the planning of scientific information systems of the future;
- the reduction of administrative and legal barriers to the flow of scientific information between and among countries.

UNISIST Programme aims are being pursued through concrete Unesco activities - generally of conceptual, normative or operational (technical assistance) nature - within the framework of the Unesco General Information Programme.* The overall programme objectives for 1977-82 activities have been defined as follows:

- promote the formulation of information policies and plans (national, regional and global);
- promote the establishment of norms and their dissemination;
- assist in the development of information infrastructures and specialized international information systems;
- promote the training and education of professionals and users of information.

The detailed target for each objective as well as the scope of envisaged programme action may be found in the medium-term plan of Unesco (2). In developing activities in the information field, Unesco is intensifying cooperation with Member governments for improvement of infrastructure and international exchange of information. Complementary cooperation is being developed with international organizations within the United Nations family in which UNISIST is recognized as providing the

---

*The 19th Unesco General Conference in November 1976 decided (Resolution 5.1) on the establishment of a single Unesco General Information Programme, effective in January 1977. The merged General Information Programme is carried out under the authority of the Director-General of Unesco on the basis of plans adopted by the General Conference, which also established an Intergovernmental Council of thirty Member States to guide the implementation process. The General Conference instructed the Council to ensure continuity in the development of the activities undertaken in the context of the UNISIST programme, and to promote the concept of overall planning of national information systems (NATIS), paying special attention to increasing the special contributions of libraries to the development of education, science, and culture, and to promoting the development of archives services. Commensurate changes have been introduced in the structure of the Unesco Secretariat, by creating a Division of the General Information Programme which combines the former Division of Scientific and Technological Information and Documentation with the personnel and resources devoted to documentation, libraries, and archives within the Culture and Communication Sector (part of the former Division of Documentation, Libraries and Archives, DBA). The new division is placed under the authority of the Assistant Director General for Studies and Programming and is thus intersectoral in nature.

conceptual framework for development of specialized information systems and programmes, and ties are being continued and strengthened with the international non-governmental organizations and professional communities with an interest in information transfer.

The objectives and principles of action of the Unesco General Information Programme perhaps can be best shown in the present context through the example of numerical data activities.

NUMERICAL DATA ACTIVITIES WITHIN THE GENERAL INFORMATION PROGRAMME

Policies and plans. The importance of numerical data to information users was recognized from the inception of the UNISIST Programme, e. g., in Recommendation 10 of the Unesco/ICSU study report:

> *"The collection, critical evaluation, organization, and dissemination of numerical data, a field in which CODATA (the Committee on Data for Science and Technology of ICSU) represents the interests of the international scientific unions, is functionally closely related to the processing of published literature, and must be provided for in any future network of information services in accordance with UNISIST principles. Special attention should be paid to the development of networking capability among numerical data centres, and to the functional relationship of such centres with the bibliographically oriented network."*

As a first step in the implementation of this recommendation, Unesco commissioned a *Study on the problems of accessibility and dissemination of data for science and technology* (3), completed by CODATA in 1974; this study provided an overview of the world situation in the field of evaluation, compilation, and dissemination of reliable data and recommendations for future action, paying special attention to the needs of developing countries. The principal recommendations of the study were subsequently incorporated into the long-term plan programme of CODATA and, in turn, certain aspects of this programme have been selected by the intergovernmental bodies supervising the Unesco General Information Programme (the Intergovernmental Council, its predecessor the UNISIST Steering Committee, and the Unesco General Conference) as a basis for a long-term UNISIST action in the field of numerical data, to be implemented in close cooperation with CODATA.

Norms and their dissemination. *A Guide for the presentation in the primary literature of numerical data derived from experiments* (4), prepared under contract by CODATA, was published as a UNISIST document in 1973; it has had wide international impact in standardized reporting of numerical data and has been forwarded to ISO (the International Organization for Standardization) for possible transformation into an international standard.

A joint Working Group of ICSU/AB (ICSU Abstracting Board) and CODATA completed a study of the problem of flagging and tagging of numerical data (5) under Unesco contract. The further development of machine-readable methods of identifying data is seen as an important aspect within the Unesco programme; the detailed recommendations of the ICSU AB/CODATA report are under active consideration by the respective ICSU Committees.

A draft international exchange format for numerical data is being developed by CODATA under contract to Unesco. It is expected that preliminary testing will be completed during 1979 and that the format will then be available for formal consideration by the international community.

A fourth aspect of standardization being considered for future action is that of copyright and related problems in the dissemination of data.

Development of infrastructures and international information. The General Information Programme is providing advice and assistance to Member States in the development of data collection and data centres in the context of national information systems, with special emphasis on the needs of developing countries. At the international level Unesco is providing initial support for activities of the World Data Referral Centre (WDRC) located within the CODATA Secretariat in Paris, and is encouraging the development of a world referral network encompassing WDRC, international and regional organizations concerned with data and national information systems.

The major initial effort of WDRC is the preparation of four directories for use in data referral (written data sources, institutional sources, individual experts, and information centres participating in the data referral network); the directories are planned to be published in 1980.

A brochure describing the activities of WDRC has been prepared by CODATA and Unesco, as has a brochure entitled "Obtaining Reliable Data" which is intended to inform information users on the scope of available numerical data (6).

Training. A long-term programme for improving awareness of scientists and engineers in the handling of experimental data is being developed in collaboration with CODATA and Unesco. Short international training courses in this field have been sponsored by CODATA and Unesco in Yugoslavia (1976) and Poland (1977) with the support of the national authorities in these countries (7). It is anticipated that experience gained in this context will serve as a basis for an extensive series of national courses in each of the geographic regions of the world; the first such effort has been a workshop organized in India in 1978.

Training of information professionals in techniques for handling numerical data is also being encouraged through the Unesco programme, and curricula are being developed for training disseminators of data and teachers in the library/information field. Short international courses of this type are tentatively envisaged in 1979-80, as is the preparation of material to help integrate data-handling techniques into existing training programmes in the information field.

REFERENCES

1. *UNISIST-Study report on the feasibility of a World Science Information System*, Unesco, Paris (1971).

2. *Medium-Term Plan (1977-82)* Unesco, Paris (1977).

3. *UNISIST, Study on the problems of accessibility and dissemination of data for science and technology*, Unesco, Paris (1974).

4. *UNISIST, Guide for the presentation in the primary literature of numerical data derived from experiments*, Unesco, Paris (1973). English, French, and Spanish versions. Also available in English as a CODATA publication, ***CODATA Bulletin*** No. **9** (December 1973), in Japanese (Tokyo, Butsuri), in Swedish (Stockholm, Royal Academy of Sciences), and in Russian (Gosstandard, Moscow).

5. ICSU AB/CODATA Joint Working Group on Tagging and Flagging. "Flagging and tagging data to indicate its presence and facilitate its retrieval." ***CODATA Bulletin*** No. **19** (June 1976).

6. "Obtaining Reliable Data" (brochure) Paris, ICSU/CODATA and Unesco (1977). Available in English, French, and Spanish.

7. *International Training Courses in the Handling of Experimental Data*, CODATA, Paris (January 1978) (***CODATA Bulletin*** No. **26**).

Data Handling for Science and Technology
S.A. Rossmassler and D.G. Watson (eds.)
North-Holland Publishing Company

APPENDIX III

CODATA AND CODATA CONFERENCES

CODATA is governed by a General Assembly with delegates from sixteen member countries and fifteen ICSU Scientific Unions. The General Assembly meets biennially.

1978-1979 National Members of CODATA

Australia, Brazil, Canada, Federal Republic of Germany, France, German Democratic Republic, India, Israel, Italy, Japan, Netherlands, Poland, Sweden, United Kingdom, United States of America, Union of Soviet Socialist Republics.

1976-1977 International Union Members of CODATA

International Astronomical Union
Observatoire de Genève
1290 Sauverny (GE), SWITZERLAND

International Union of Geodesy and Geophysics
Observatoire Royal de Belgique
3, avenue Circulaire
1180 Brussels, BELGIUM

International Union of Pure and Applied Chemistry
2-3 Pound Way, Cowley Centre
Oxford OX4 3YF, UNITED KINGDOM

International Union of Pure and Applied Physics
Physics Department
Laval University
Quebec 10, G1K 7P4, CANADA

International Union of Biological Sciences
51 Boulevard de Montmorency
75016 Paris, FRANCE

International Geographical Union
United Nations University
29 Floor, Toho Seimei Bldg.,
15-1 Shibuy 2-chome Shibuya-ku
Tokyo 150, JAPAN

International Union of Crystallography
5 Abbey Square
Chester CH1 2HU, UNITED KINGDOM

International Union of Physiological Sciences
Department of Physiology

Middlesex Hospital Medical School
Windeyer Building - Cleveland Street
London W1P 6DB, UNITED KINGDOM

International Union of Theoretical and Applied Mechanics
Chalmers University of Technology
S-40220 Gothenburg, SWEDEN

International Union of Biochemistry
Biochemistry-UMED
P. O. Box 520875
Miami, Florida 33152, USA

International Union of Geological Sciences
Rijks Geol. Dienst. Postbus 379
Haarlem, NETHERLANDS

International Union of Pure and Applied Biophysics
Physiological Lab.
Downing Street
Cambridge CB2 3EG, UNITED KINGDOM

International Union of Nutritional Sciences
Institute of Clinical Nutrition
University of Gothenburg
Sahlgren's Hospital
45 Gothenburg S-413, SWEDEN

International Union of Pharmacology
Department of Pharmacology
University of Heidelberg
Im Neuenheimer Feld 366
6900 Heidelberg, F.R.G.

International Union of Immunological Sciences
Institut Pasteur
75015 Paris, FRANCE

In addition CODATA has Co-opted Delegates from

World Data Centres and

Federation of Astronomical and Geophysical Services.

A number of international bodies are Associate Organizations of CODATA.

A major feature of the total CODATA program is the series of biennial International CODATA Conferences. Unique among scientific conferences, these meetings are devoted to various aspects of the worldwide effort to supply reliable data in response to users' needs. Six International CODATA Conferences have been held since CODATA was founded -- in 1968 (Arnoldshain, Germany), 1970 (St. Andrews, Scotland, U.K.), 1972 (Le Creusot, France), 1974 (Tsakhcadzor, USSR), 1976 (Boulder, Colorado, USA), and 1978 (Palermo, Sicily). For 1980 the Conference will be held in Kyoto, Japan; for 1982 in Poland.

CODATA Conferences occupy a full working week. A full formal program is arranged for five days, while special sessions are provided for Task Groups and other units during the official meeting.

The Proceedings of each Conference are edited by the CODATA Secretariat, and published as promptly after the meeting as possible. The content of these Proceedings is highly relevant to the interests of most persons concerned with aspects of data handling. To confirm this point, and to make it easier for the reader to locate the papers he may wish, the Tables of Contents of the Proceedings of the four most recent Conferences (1972, -74, -76, -78) are reprinted below.

A further source of details concerning scientific and technical data, data handling, and the problems of accessibility and dissemination of data is the UNISIST report SC.74/WS16, cited in Chapter 1 as Reference 1. The Table of Contents of that report also is reproduced below, following those of the CODATA Conference Proceedings.

PROCEEDINGS OF THE

THIRD INTERNATIONAL CODATA CONFERENCE

Le CREUSOT, FRANCE, 26-29 June 1972

Table of Contents

Earth and Atmospheric Sciences

Biological Sciences

Astronomy and Astrophysics

Engineering Sciences

Computer Topics of Special Interest to Data Analysis Centres

Special Topics Important to Data Analysis Centre Operations

Accessibility and Dissemination of Critical Reviews and Data Compilations

Additional Presentation

PROCEEDINGS OF THE FOURTH INTERNATIONAL CODATA CONFERENCE, TSAKHCADZOR, U.S.S.R., 24–27 JUNE 1974

Table of Contents

Formulation of CODATA's Role in Meeting the Needs of the Biological Sciences

Formulation of CODATA's Role in Meeting the Needs of the Geological, Geophysical, Geographical, and Astronomical Sciences

Progress in Handling Spectroscopic Data

Progress in Handling Thermodynamic Data

PROCEEDINGS OF THE

FIFTH INTERNATIONAL CODATA CONFERENCE,

BOULDER, COLORADO, U.S.A. 28 JUNE - 1 JULY 1976

Table of Contents

Data for Technology

Data on Flavor and Aroma

Data Tagging and Related Activities

Biological Data Banking and Networking

Biological Data Compilation and Application

Data Activities Relevant to Energy Research and Development

Plenary Session

Evaluation of Data on Thermophysical Properties of Fluids

Computer Techniques in Numerical Data Handling

Computer Systems for Numerical Data Handling

PROCEEDINGS OF THE

SIXTH INTERNATIONAL CODATA CONFERENCE

SANTA FLAVIA, PALERMO, ITALY 22 - 25 MAY 1978

Table of Contents

Industrial Data and International Endeavours

Data and Computerized Information Systems

Data Evaluation and Dissemination

Spectroscopy and Computer-Aided Elucidation of Structures

---

*Papers in parentheses were presented orally in Santa Flavia, but are not included in full in the Proceedings.

---

*Papers in parentheses were presented orally in Santa Flavia, but are not included in full in the Proceedings.

Data in the Physical Sciences

Data Evaluation Methodology

---

*Papers in parentheses were presented orally in Santa Flavia, but are not included in full in the Proceedings.

Data Needs for Energy

Future Trends in CODATA

Astro- and Geosciences

Scientific Data Correlation

---

*Papers in parentheses were presented orally in Santa Flavia, but are not included in full in the Proceedings.

Data Handling and Computer Systems

Data Processing Methodology

---

*Papers in parentheses were presented orally in Santa Flavia, but are not included in full in the Proceedings.

UNISIST REPORT SC.74/WS/16,

"STUDY ON THE PROBLEMS OF ACCESSIBILITY

AND DISSEMINATION OF DATA FOR SCIENCE AND TECHNOLOGY"

Table of Contents

Use of Computer and Telecommunication Technologies in Data Services

Access to Data by Traditional Channels

Data Evaluation, Dissemination and Referral Services

Prerequisites for Design and Implementation of Data Dissemination Services, with Special Reference to Developing Countries

INDEX

The Index which follows is intended to help the reader find substantive discussion or information in the text relating to the index terms. Organizations, countries, and important concepts are merged in a single list of terms. Acronyms are listed alphabetically with the corresponding full names following in parentheses. The references are to page numbers. If the text deals with a concept on several consecutive pages, only the first page may be entered. References to material in the Appendices do not carry page numbers, since the format of the Appendices makes it easy to locate the items of interest.